Congressional Research Service

Navy DDG-51 and DDG-1000 Destroyer Programs: Background and Issues for Congress

Ronald O'Rourke
Specialist in Naval Affairs

October 18, 2012

Congressional Research Service

7-5700

www.crs.gov

RL32109

CRS Report for Congress ————
Prepared for Members and Committees of Congress

Summary

The Navy's FY2013 budget submission calls for procuring nine Arleigh Burke (DDG-51) class destroyers in FY2013-FY2017, in annual quantities of 2-1-2-2-2. The five DDG-51s scheduled for procurement in FY2013-FY2015, and one of the two scheduled for procurement in FY2016, are to be of the current Flight IIA design. The Navy wants to begin procuring a new version of the DDG-51 design, called the Flight III design, starting with the second of the two ships scheduled for procurement in FY2016. The two DDG-51s scheduled for procurement in FY2017 are also to be of the Flight III design. The Flight III design is to feature a new and more capable radar called the Air and Missile Defense Radar (AMDR). The Navy this year is requesting congressional approval to use a multiyear procurement (MYP) arrangement for the nine DDG-51s scheduled for procurement in FY2013-FY2017.

The Navy's proposed FY2013 budget requests $3,048.6 million to complete the procurement funding for the two DDG-51s scheduled for procurement in FY2013. The Navy estimates the total procurement cost of these ships at $3,149.4 million, and the ships have received $100.7 million in prior-year advance procurement (AP) funding. The FY2013 budget also requests $466.3 million in AP funding for DDG-51s to be procured in future fiscal years. Much of this AP funding is for Economic Order Quantity (EOQ) procurement of selected components of the nine DDG-51s to be procured under the proposed FY2013-FY2017 MYP arrangement. The Navy's proposed FY2013 budget also requests $669.2 million in procurement funding to help complete procurement costs for three Zumwalt (DDG-1000) class destroyers procured in FY2007-FY2009, and $223.6 million in research and development funding for the AMDR.

A Government Accountability Office (GAO) report released on January 24, 2012, discusses several potential oversight issues for Congress regarding the Navy's plans for procuring DDG-51s, particularly the Flight III version. Some of these issues were first raised in this CRS report; the GAO report developed these issues at length and added some additional issues. Potential FY2013 issues for Congress concerning destroyer procurement include the following:

- whether actions should be taken to mitigate the projected shortfall in cruisers and destroyers;

- whether to approve the Navy's request for a DDG-51 MYP arrangement beginning in FY2013, and if so, whether it should include Flight III DDG-51s;

- the possibility of adding a 10th ship to the proposed DDG-51 MYP arrangement;

- whether there is an adequate analytical basis for procuring Flight III DDG-51s in lieu of the previously planned CG(X) cruiser;

- whether the Flight III DDG-51 would have sufficient air and missile capability to adequately perform future air and missile defense missions;

- cost, schedule, and technical risk in the Flight III DDG-51 program;

- whether the Flight III DDG-51 design would have sufficient growth margin for a projected 35- or 40-year service life; and

- schedule risk for recently procured Flight IIA DDG-51s.

Contents

Figures

Tables

Appendixes

Contacts

Introduction

This report presents background information and potential oversight issues for Congress on the Navy's Arleigh Burke (DDG-51) and Zumwalt (DDG-1000) class destroyer programs. The Navy's proposed FY2013 budget requests funding for the procurement of two DDG-51s. The Navy this year is also requesting congressional approval to use a multiyear procurement (MYP) arrangement for the nine DDG-51s scheduled for procurement in FY2013-FY2017. Decisions that Congress makes concerning these programs could substantially affect Navy capabilities and funding requirements, and the U.S. shipbuilding industrial base.

Background

DDG-51 Program

General

The DDG-51 program was initiated in the late 1970s.[1] The DDG-51 (**Figure 1**) is a multi-mission destroyer with an emphasis on air defense (which the Navy refers as anti-air warfare, or AAW) and blue-water (mid-ocean) operations. DDG-51s, like the Navy's 22 Ticonderoga (CG-47) class cruisers,[2] are equipped with the Aegis combat system, an integrated ship combat system named for the mythological shield that defended Zeus. CG-47s and DDG-51s consequently are often referred to as Aegis cruisers and Aegis destroyers, respectively, or collectively as Aegis ships. The Aegis system has been updated several times over the years. Existing DDG-51s (and also some CG-47s) are being modified to receive an additional capability for ballistic missile defense (BMD) operations.[3]

The first DDG-51 was procured in FY1985. A total of 66 have been procured through FY2012, including 62 in FY1985-FY2005, none during the four-year period FY2006-FY2009, one in FY2010, two in FY2011, and one more in FY2012. The first DDG-51 entered service in 1991, and a total of 61 were in service as of the end of FY2011. Of the 66 DDG-51s procured through FY2012, General Dynamics Bath Iron Works (GD/BIW) of Bath, ME, is the builder of 36, and Ingalls Shipbuilding of Pascagoula, MS, a division of Huntington Ingalls Industries (HII), is the builder of 30.

The DDG-51 design has been modified over time. The first 28 DDG-51s (i.e., DDGs 51 through 78) are called Flight I/II DDG-51s. Subsequent ships in the class (i.e., DDGs 79 and higher) are

---

[1] The program was initiated with the aim of developing a surface combatant to replace older destroyers and cruisers that were projected to retire in the 1990s. The DDG-51 was conceived as an affordable complement to the Navy's Ticonderoga (CG-47) class Aegis cruisers.

[2] A total of 27 CG-47s were procured for the Navy between FY1978 and FY1988; the ships entered service between 1983 and 1994. The first five, which were built to an earlier technical standard, were judged by the Navy to be too expensive to modernize and were removed from service in 2004-2005.

[3] The modification for BMD operations includes, among other things, the addition of a new software program for the Aegis combat system and the arming of the ship with the SM-3, a version of the Navy's Standard Missile that is designed for BMD operations. For more on Navy BMD programs, CRS Report RL33745, *Navy Aegis Ballistic Missile Defense (BMD) Program: Background and Issues for Congress*, by Ronald O'Rourke.

---

referred to as Flight IIA DDG-51s. The Flight IIA design, first procured in FY1994, implemented a significant design change that included, among other things, the addition of a helicopter hangar. The Flight IIA design has a full load displacement of about 9,500 tons, which is similar to that of the CG-47.

Figure 1. DDG-51 Class Destroyer

Source: Navy file photograph accessed online October 18, 2012, at http://www.navy.mil/view_image.asp?id= 134605.

The Navy's FY2013 30-year (FY2013-FY2042) shipbuilding plan, like the FY2012 30-year shipbuilding plan, assumes a 35-year service life for Flight I/II DDG-51s and a 40-year service life for Flight IIA DDG-51s. The Navy is implementing a program for modernizing all DDG-51s (and CG-47s) so as to maintain their mission and cost effectiveness out to the end of their projected service lives.[4]

Older CRS reports provide additional historical and background information on the DDG-51 program.[5]

---

[4] For more on this program, see CRS Report RS22595, *Navy Aegis Cruiser and Destroyer Modernization: Background and Issues for Congress*, by Ronald O'Rourke.

[5] See CRS Report 94-343, *Navy DDG-51 Destroyer Procurement Rate: Issues and Options for Congress*, by Ronald O'Rourke (April 25, 1994; out of print and available directly from the author), and CRS Report 80-205, *The Navy's Proposed Arleigh Burke (DDG-51) Class Guided Missile Destroyer Program: A Comparison With An Equal-Cost Force Of Ticonderoga (CG-47) Class Guided Missile Destroyers*, by Ronald O'Rourke (November 21, 1984; out of print and available directly from the author).

Resumption of Flight IIA DDG-51 Procurement in FY2010

The Navy in July 2008 announced that it wanted to end procurement of DDG-1000 destroyers (see "DDG-1000 Program" below) and resume procurement of Flight IIA DDG-51s. The announcement represented a major change in Navy planning: prior to the announcement, the Navy for years had strongly supported ending DDG-51 procurement permanently in FY2005 and proceeding with procurement of DDG-1000 destroyers.[6] The Navy's FY2010 budget, submitted in May 2009, reflected the Navy's July 2008 change in plans: the budget proposed truncating DDG-1000 procurement to the three ships that had been procured in FY2007 and FY2009, and resuming procurement of Flight IIA DDG-51s. Congress, as part of its action on the FY2010 defense budget, supported the proposal.[7]

Procurement of First Flight III DDG-51 Planned for FY2016

The Navy's FY2011 budget, submitted in February 2010, proposed another major change in Navy plans—terminating a planned cruiser called the CG(X) in favor of procuring an improved version of the DDG-51 called the Flight III version.[8] Rather than starting to procure CG(X)s around FY2017, Navy plans call for procuring the first Flight III DDG-51 in FY2016.

Compared to the Flight IIA DDG-51 design, the Flight III design is to feature a new and more capable radar called the Air and Missile Defense Radar (AMDR). The version of the AMDR to be carried by the Flight III DDG-51 is smaller and less powerful than the one envisaged for the CG(X): the Flight III DDG-51's AMDR is to have a diameter of 12 or 14 feet, while the AMDR envisaged for the CG(X) would have had a substantially larger diameter.[9]

---

[6] The Navy announced this change in its plans at a July 31, 2008, hearing before the Seapower and Expeditionary Forces subcommittee of the House Armed Services Committee. In explaining their proposed change in plans, Navy officials cited a reassessment of threats that Navy forces are likely to face in coming years. As a result of this reassessment, Navy officials stated, the service decided that destroyer procurement over the next several years should emphasize three mission capabilities—area-defense AAW, BMD, and open-ocean ASW. Navy officials also stated that they want to maximize the number of destroyers that can be procured over the next several years within budget constraints. Navy officials stated that DDG-51s can provide the area-defense AAW, BMD, and open-ocean ASW capabilities that the Navy wants to emphasize, and that while the DDG-1000 design could also be configured to provide these capabilities, the Navy could procure more DDG-51s than reconfigured DDG-1000s over the next several years for the same total amount of funding. In addition, the Navy by 2008-2009 no longer appeared committed to the idea of reusing the DDG-1000 hull as the basis for the Navy's planned CG(X) cruiser. If the Navy had remained committed to that idea, it might have served as a reason for continuing DDG-1000 procurement.

[7] The FY2010 budget funded the procurement of one DDG-51, provided advance procurement funding for two DDG-51s the Navy wants to procure in FY2011, completed the procurement funding for the third DDG-1000 (which was authorized but only partially funded in FY2009), and provided no funding for procuring additional DDG-1000s.

[8] The Navy stated that its desire to terminate the CG(X) program was "driven by affordability considerations." (Department of the Navy, Office of Budget, *Highlights of the Department of the Navy FY 2011 Budget*, February 2010, p. 5-7.) For more on the CG(X) program and its termination in favor of procuring Flight III DDG-51s, see **Appendix B**.

[9] Government Accountability Office, *Arleigh Burke Destroyers[:] Additional Analysis and Oversight Required to Support the Navy's Future Surface Combatant Plans*, GAO-12-113, January 2012, pp. 31 and 42. See also Zachary M. Peterson, "DDG-51 With Enhanced Radar in FY-16, Design Work To Begin Soon," *Inside the Navy*, February 8, 2010; Amy Butler, "STSS Prompts Shift in CG(X) Plans," *Aerospace Daily & Defense Report*, December 11, 2010: 1-2; "[Interview With] Vice Adm. Barry McCullough," *Defense News*, November 9, 2009: 38.

The written testimony of the Chief of Naval Operations (CNO) before the House Armed Services Committee on February 16, 2012, and before the Defense subcommittee of the House Appropriations Committee on March 1, 2012, stated that the Flight III design would use an all-electric propulsion system, in contrast to the mechanical propulsion system used on the Flight IIA design and other Navy surface combatants. (See, for example, Statement of Admiral (continued...)

On July 24, 2012, Frank Kendall, the Under Secretary of Defense for Acquisition, Technology and Logistics (or USD ATL—the acquisition executive for the Department of Defense), designated the DDG-51 program as an Acquisition Category (ACAT) 1D program, meaning that he (rather than the Secretary of the Navy or the Navy's acquisition executive) will act as the Milestone Decision Authority (MDA) for the DDG-51 program.

As mentioned earlier, the two DDG-51s that the Navy wants to procure in FY2016 include the final Flight IIA DDG-51 and the first Flight III DDG-51. The combined cost for these two ships shown in the Navy's FY2013 budget submission suggests that the Navy estimates the procurement cost of the first Flight III DDG-51 at roughly $2.3 billion. The FY2013 budget estimates that the two Flight III DDG-51s scheduled for procurement in FY2017 would cost an average of $2.12 billion each.

Request for Multiyear Procurement (MYP) in FY2013-FY2017

General

The Navy this year is requesting congressional approval to use a multiyear procurement (MYP) arrangement[10] for the nine DDG-51s scheduled for procurement in FY2013-FY2017. This MYP arrangement would include the final six Flight IIA DDG-51s in FY2013-FY2016 and the first three Flight III DDG-51s in FY2016-FY2017. It would be the third MYP arrangement for the DDG-51 program: the program used an MYP arrangement to procure 13 ships (all Flight IIA ships) in FY1998-FY2001, and another MYP arrangement to procure 11 ships (again all Flight IIA ships) in FY2002-FY2005.

The Navy estimates that procuring the nine DDG-51s scheduled for procurement in FY2013-FY2017 under an MYP arrangement would reduce their combined procurement cost by $1,538.1 million in then-year dollars, or about 8.7%, compared to procuring these nine ships with separate annual contracts. The estimated savings when calculated in real (i.e., inflation-adjusted) terms are $1,400.1 million in constant FY2012 dollars, or about 8.5%. The estimated savings when calculated on a net present value (NPV) basis are $1,308.1 million, or about 8.4%.[11]

(...continued)

Jonathan Greenert, Chief of Naval Operations, Before the House Armed Services Committee [hearing] on FY 2013 Department of the Navy Posture, February 16, 2012, which stated on page 10: "Our Lewis and Clark class supply ships now employ an all-electric propulsion system, as will our new Zumwalt and Flight III Arleigh Burke class destroyers (DDG).") The written testimony of the CNO before the Defense subcommittee of the Senate Appropriations Committee on March 7, 2012, and before the Senate Armed Services Committee on March 15, 2012, omitted the reference to the Flight III DDG-51 being equipped with an all-electric propulsion system. In response to a question from CRS about the change in the testimony, the Navy informed CRS on March 15, 2012, that the statement in the earlier testimony was an error, and that the Flight III DDG-51 will likely not be equipped with an all-electric propulsion system.

[10] For an explanation of MYP, see CRS Report R41909, *Multiyear Procurement (MYP) and Block Buy Contracting in Defense Acquisition: Background and Issues for Congress*, by Ronald O'Rourke and Moshe Schwartz.

[11] Department of the Navy, *Department of the Navy Fiscal year (FY) 2013 Budget Estimates, Justification of Estimates, Shipbuilding and Conversion, Navy*, February 2012, Exhibit MYP-4 Present Value Analysis (Navy) (MYP, page 9 of 9), PDF page 155 of 246.

Possibility of Including a 10ᵗʰ Ship

At a March 29, 2012, hearing on Navy shipbuilding programs before the Seapower and Projection Forces subcommittee of the House Armed Services Committee, Sean Stackley, the Assistant Secretary of the Navy for Research, Development, and Acquisition (i.e., the Navy's acquisition executive), stated the following in response to a question about the FY2013 budget's deferral to FY2016 of a second DDG-51 that was previously programmed for FY2014—a deferral that Navy officials have testified was made in response to FY2014 becoming a tight budget year for the Navy:

> I'd like to address the question regarding the second destroyer in 2014. A couple of important facts: First, the—we restarted DDG-51 construction in—in [FY]2010 and we've got four ships under contract, and a result of the four ships that we've placed under contract is we have prior year savings in this program that are—work in our favor when we consider future procurement for the [DDG-]51s.
>
> We also have a unique situation where we've got competition on this program—two builders building the 51s, and the competition has been healthy with both builders. We also have a very significant cost associated with government-furnished equipment, so not only did we restart construction at the shipyards, we also restarted manufacturing lines at our weapon systems providers.
>
> So in this process we were able to restart 51s virtually without skipping a beat, and we're seeing the continued learning curve that we left off on back with the [FY]2005 procurement. So when we march into this third multiyear for the 51s we're looking to capitalize on the same types of savings that we saw prior, and our top line, again, allowed for nine ships to be budgeted, but when we go out with this procurement we're going to go out with a procurement that enables the procurement of 10 ships, where that 10ᵗʰ ship would be the second—potentially the second ship in [FY]2014 if we're able to achieve the savings that we're targeting across this multiyear between the shipbuilders in competition as well as the combat systems providers as well as all of the other support and engineering associated with this program.
>
> So we want to leverage the strong learning, we want to leverage the strong industrial base, we want to leverage the competition to get to what we need in terms of both affordability and force structure, and I think we have a pretty good shot at it.[12]

DDG-1000 Program

The DDG-1000 program was initiated in the early 1990s.[13] The DDG-1000 is a multi-mission destroyer with an emphasis on naval surface fire support (NSFS) and operations in littoral (i.e., near-shore) waters. The DDG-1000 is intended to replace, in a technologically more modern form, the large-caliber naval gun fire capability that the Navy lost when it retired its Iowa-class battleships in the early 1990s,[14] to improve the Navy's general capabilities for operating in

---

[12] Source: Transcript of hearing. See also Megan Eckstein, "Navy Looking Into Feasibility Of Procuring 10ᵗʰ DDG In Multiyear Contract," *Inside the Navy*, April 2, 2012.

[13] The program was originally designated DD-21, which meant destroyer for the 21ˢᵗ Century. In November 2001, the program was restructured and renamed DD(X), meaning a destroyer whose design was in development. In April 2006, the program's name was changed again, to DDG-1000, meaning a guided missile destroyer with the hull number 1000.

[14] The Navy in the 1980s reactivated and modernized four Iowa (BB-61) class battleships that were originally built (continued...)

---

defended littoral waters, and to introduce several new technologies that would be available for use on future Navy ships. The DDG-1000 was also intended to serve as the basis for the Navy's now-canceled CG(X) cruiser.

The DDG-1000 is to have a reduced-size crew of 142 sailors (compared to roughly 300 on the Navy's Aegis destroyers and cruisers) so as to reduce its operating and support (O&S) costs. The ship incorporates a significant number of new technologies, including an integrated electric-drive propulsion system[15] and automation technologies enabling its reduced-sized crew.

With an estimated full load displacement of 15,482 tons, the DDG-1000 design is roughly 63% larger than the Navy's current 9,500-ton Aegis cruisers and destroyers, and larger than any Navy destroyer or cruiser since the nuclear-powered cruiser *Long Beach* (CGN-9), which was procured in FY1957.

The first two DDG-1000s were procured in FY2007 and split-funded (i.e., funded with two-year incremental funding) in FY2007-FY2008; the Navy's FY2013 budget submission estimates their combined procurement cost at $7,795.2 million. The third DDG-1000 was procured in FY2009 and split-funded in FY2009-FY2010; the Navy's FY2013 budget submission estimates its procurement cost at $3,674.9 million.

The estimated combined procurement cost for all three ships in the FY2013 budget submission—$11,470.1 million—is $161.3 million, or about 1.4%, higher than the figure of $11,308.8 million shown in the FY2012 budget, which in turn was $1,315.5 million, or about 13.2%, higher than the figure of $9,993.3 million shown in the FY2011 budget. The Navy stated the following regarding the 13.2% increase between the FY2011 and FY2012 budgets:

> The increase in end cost between PB11 [the President's proposed budget for FY2011] and PB12 is $1,315.5M [million] for all three ships. $211.7M was added via Special Transfer Authority to address ship cost increases as a result of market fluctuations and rate adjustments on DDG 1000 and DDG 1001. $270M was added due to updated estimates for class services and technical support that would have been spread throughout ships 4-7. The remaining $833.8M was added to fund to the most recent estimate of construction and mission systems equipment (MSE) procurement costs.[16]

All three ships are to be built at GD/BIW, with some portions of each ship being built by Ingalls Shipbuilding for delivery to GD/BIW. Raytheon is the prime contractor for the DDG-1000's combat system (its collection of sensors, computers, related software, displays, and weapon launchers). The Navy awarded GD/BIW the contract for the construction of the second and third DDG-1000s on September 15, 2011.[17]

(...continued)

during World War II. The ships reentered service between 1982 and 1988 and were removed from service between 1990 and 1992.

[15] For more on integrated electric-drive technology, see CRS Report RL30622, *Electric-Drive Propulsion for U.S. Navy Ships: Background and Issues for Congress*, by Ronald O'Rourke.

[16] Navy information paper dated May 2, 2011, and provided by the Navy to CRS on June 9, 2011. See also John M. Donnelly, "Building Budgets On Shaky Moorings," *CQ Weekly*, September 26, 2011: 1974-1976.

[17] See, for example, Mike McCarthy, "Navy Awards Contract for DDG-1000s," *Defense Daily*, September 16, 2011: 3-4.

For additional background information on the DDG-1000 program, see **Appendix A**.

Projected Shortfall in Cruisers and Destroyers

The Navy's FY2013 30-year (FY2013-FY2042) 30-year shipbuilding plan, submitted to Congress on March 28, 2012, reduces the Navy's cruiser-destroyer force-level objective from 94 ships to about 90 ships.[18]

The FY2013 30-year shipbuilding plan does not contain enough destroyers to maintain a force of 90 cruisers and destroyers consistently over the long run. As shown in **Table 1**, the Navy projects that implementing the FY2013 30-year shipbuilding plan would result in a cruiser-destroyer force that remains below 90 ships every year in the 30-year period except FY2027, and that bottoms out in FY2014-FY2015 and again in FY2034 at 78 ships—12 ships, or about 13.3% below the required figure of about 90 ships.

The projected cruiser-destroyer shortfall under the FY2013 30-year shipbuilding plan is not as deep at its lowest point as the projected shortfall under the FY2012 30-year shipbuilding plan. Under the FY2012 30-year plan, the cruiser-destroyer force was projected to reach a minimum of 68 ships (i.e., 26 ships, or about 28%, below the then-required figure of 94 ships) in FY2034, and remain 16 or more ships below the 94-ship figure through the end of the 30-year period.

The projected cruiser-destroyer shortfalls is less deep under the FY2013 30-year plan than under the FY2012 30-year plan in part because the FY2013 30-year plan reduces the cruiser-destroyer force-level goal from 94 ships to about 90 ships, and in part because the FY2013 30-year shipbuilding plan includes a total of 70 destroyers—18 more than were included in the FY2012 30-year shipbuilding plan. Of the 18 additional destroyers in the FY2013 plan, 15 occur in the final 20 years of the plan.

Although the projected cruiser-destroyer shortfall is less deep under the FY2013 30-year plan than it was under the FY2012 30-year plan, it remains the largest projected shortfall of any ship category in the 30-year plan.

The figures shown in **Table 1** reflect a Navy cost-saving proposal in the FY2013 budget to retire seven Aegis cruisers in FY2013 and FY2014, more than a decade before the ends of their 35-year expected service lives in FY2026-FY2029.

---

[18] This represents the second change in the cruiser-destroyer force-level goal in a year. In an April 2011 report to Congress on naval force structure and ballistic missile defense (U.S. Navy, Office of the Chief of Naval Operations, Director of Strategy and Policy (N51), *Report to Congress On Naval Force Structure and Missile Defense*, April 2011, 12 pp.), the Navy increased the cruiser-destroyer force-level goal from 88 ships to 94 ships. For more on Navy ship force-level goals, see CRS Report RL32665, *Navy Force Structure and Shipbuilding Plans: Background and Issues for Congress*, by Ronald O'Rourke.

Table 1. Projected Cruiser-Destroyer Shortfall

As shown in Navy's FY2013 30-Year Shipbuilding Plan

Fiscal year	Projected number of cruisers and destroyers	Shortfall relative to 90-ship goal, shown as a negative	
		Number of ships	Percent
13	80	-10	-11%
14	78	-12	-13%
15	78	-12	-13%
16	80	-10	-11%
17	82	-8	-9%
18	84	-6	-7%
19	86	-4	-4%
20	87	-3	-3%
21	88	-2	-2%
22	87	-3	-3%
23	89	-1	-1%
24	89	-1	-1%
25	88	-2	-2%
26	89	-1	-1%
27	90		
28	89	-1	-1%
29	87	-3	-3%
30	85	-5	-6%
31	81	-9	-10%
32	80	-10	-11%
33	79	-11	-12%
34	78	-12	-13%
35	80	-10	-11%
36	82	-8	-9%
37	84	-6	-7%
38	86	-4	-4%
39	88	-2	-2%
40	88	-2	-2%
41	89	-1	-1%
42	88	-2	-2%

Source: Table prepared by CRS based on Navy's FY2013-FY2042 30-year shipbuilding plan. Percentage figures rounded to nearest percent.

Surface Combatant Construction Industrial Base

All cruisers, destroyers, and frigates procured since FY1985 have been built at General Dynamics' Bath Iron Works (GD/BIW) shipyard of Bath, ME, and Ingalls Shipbuilding of Pascagoula, MS, a division of Huntington Ingalls Industries (HII).[19] Both yards have long histories of building larger surface combatants. Construction of Navy surface combatants in recent years has accounted for virtually all of GD/BIW's ship-construction work and for a significant share of Ingalls' ship-construction work. (Ingalls also builds amphibious ships for the Navy.) Navy surface combatants are overhauled, repaired, and modernized at GD/BIW, Ingalls, other private-sector U.S. shipyards, and government-operated naval shipyards (NSYs).

Lockheed Martin and Raytheon are generally considered the two leading Navy surface combatant radar makers and combat system integrators. Northrop Grumman is a third potential maker of Navy surface combatant radars. Lockheed is the lead contractor for the DDG-51 combat system (the Aegis system), while Raytheon is the lead contractor for the DDG-1000 combat system, the core of which is called the Total Ship Computing Environment Infrastructure (TSCE-I). Lockheed has a share of the DDG-1000 combat system, and Raytheon has a share of the DDG-51 combat system. Lockheed, Raytheon, and Northrop are competing to be the maker of the AMDR to be carried by the Flight III DDG-51.

The surface combatant construction industrial base also includes hundreds of additional firms that supply materials and components. The financial health of Navy shipbuilding supplier firms has been a matter of concern in recent years, particularly since some of them are the sole sources for what they make for Navy surface combatants.

FY2013 Funding Request

The Navy's proposed FY2013 budget requests $3,048.6 million to complete the procurement funding for the two DDG-51s scheduled for procurement in FY2013. The Navy estimates the total procurement cost of these ships at $3,149.4 million, and the ships have received $100.7 million in prior-year advance procurement (AP) funding. The FY2013 budget also requests $466.3 million in AP funding for DDG-51s to be procured in future fiscal years. Much of this AP funding is for Economic Order Quantity (EOQ) procurement of selected components of the nine DDG-51s to be procured under the proposed FY2013-FY2017 MYP arrangement. The Navy's proposed FY2013 budget also requests $669.2 million in procurement funding to help complete procurement costs for three Zumwalt (DDG-1000) class destroyers procured in FY2007-FY2009, and $223.6 million in research and development funding for the AMDR. The funding request for the AMDR is contained in the Navy's research and development account in Project 3186 ("Air and Missile Defense Radar") of Program Element (PE) 0604501N ("Advanced Above Water Sensors").

---

[19] HII was previously owned by Northrop Grumman, during which time it was known as Northrop Grumman Shipbuilding.

Issues for Congress

The Navy's plan for procuring Flight IIA DDG-51s followed by Flight III DDG-51s poses several potential oversight issues for Congress. A January 2012 Government Accountability Office (GAO) report discusses a number of these issues.[20] Some of the issues discussed in the GAO report were first raised in this CRS report; the GAO report developed these issues at length and added some additional issues.

Potential FY2013 issues for Congress concerning destroyer procurement include the following:

- whether actions should be taken to mitigate the projected shortfall in cruisers and destroyers;

- whether to approve the Navy's request for a DDG-51 MYP arrangement beginning in FY2013, and if so, whether it should include Flight III DDG-51s;

- the possibility of adding a 10th ship to the proposed DDG-51 MYP arrangement;

- whether there is an adequate analytical basis for procuring Flight III DDG-51s in lieu of the previously planned CG(X) cruiser;

- whether the Flight III DDG-51 would have sufficient air and missile capability to adequately perform future air and missile defense missions;

- cost, schedule, and technical risk in the Flight III DDG-51 program;

- whether the Flight III DDG-51 design would have sufficient growth margin for a projected 35- or 40-year service life; and

- schedule risk for recently procured Flight IIA DDG-51s.

Each of these issues is discussed below.

Projected Cruiser-Destroyer Shortfall

One issue for Congress is whether actions should be taken to mitigate the projected shortfall in cruisers and destroyers shown in **Table 1**. Options for mitigating this projected shortfall include the following:

- keeping in service some of all of the seven Aegis cruisers proposed for early retirement in FY2013 and FY2014;

- adding DDG-51s to the Navy's shipbuilding plan; and

- extending the service lives of some Flight I/II DDG-51s to 40 or 45 years (i.e., 5 or 10 years beyond their currently planned 35-year service lives).

These options are not mutually exclusive; they could be applied in combination.

---

[20] Government Accountability Office, *Arleigh Burke Destroyers[:] Additional Analysis and Oversight Required to Support the Navy's Future Surface Combatant Plans*, GAO-12-113, January 2012, 64 pp.

Keeping in service some or all of the seven Aegis cruisers proposed for early retirement in FY2013 and FY2014 would increase the size of the cruiser-destroyer force by up to seven ships between FY2013-FY2014 and the late 2020s. The Navy has testified that keeping these seven ships in service would cost about $4.1 billion in FY2013-FY2017, and additional funding each year after FY2017. The figure of about $4.1 billion for FY2013-FY2017 includes costs for conducting maintenance and modernization work on the ships during those years, for operating the ships during those years (including crew costs), and for procuring, crewing, and operating during those years helicopters that would be embarked on the ships.[21]

Extending the service lives of Flight I/II DDG-51s could require increasing, perhaps soon, funding levels for the maintenance of these ships, to help ensure they would remain in good enough shape to eventually have their lives extended for another 5 or 10 years. This additional maintenance funding would be on top of funding that the Navy has already programmed to help ensure that these ships can remain in service to the end of their currently planned 35-year lives. The potential need to increase maintenance funding soon could make the question of whether to extend the lives of these ships a potentially near-term issue for policymakers.

A January 16, 2012, press report stated:

> The Navy will take a close look at a looming cruiser and destroyer gap over the next several budget cycles to see how the problem might be mitigated, Vice Adm. Terry Blake, deputy chief of naval operations for integration of capabilities and resources (N8), said last....
>
> He said the Navy would consider a range of options, including extending the service lives of vessels and implementing rotational crewing.
>
> "We have highlighted the problem," he said. "We're going to have to have a deliberate discussion over the next several POMs [program objective memoranda] to deal with that issue...."[22]

Request for Multiyear Procurement (MYP) in FY2013-FY2017

Another issue for Congress is whether to approve the Navy's request for a DDG-51 MYP arrangement beginning in FY2013, and if so, whether it should include Flight III DDG-51s. As mentioned earlier, the Navy this year is requesting congressional approval to use a multiyear procurement (MYP) arrangement for the nine DDG-51s scheduled for procurement in FY2013-FY2017. This MYP arrangement would include the final six Flight IIA DDG-51s in FY2013-FY2016 and the first three Flight III DDG-51s in FY2016-FY2017.

In assessing whether to approve the Navy's request for a DDG-51 MYP arrangement beginning in FY2013, and if so, whether it should include Flight III DDG-51s, Congress may consider various factors, including the following three:

- the potential for additional reductions to planned defense spending levels;

---

[21] Source: Transcript of spoken testimony of Vice Admiral William Burke, Deputy Chief of Naval Operations, Fleet Readiness and Logistics, before the Readiness subcommittee of the House Armed Services Committee, March 22, 2012.

[22] Megan Eckstein and Dan Taylor, "Blake: Navy To Examine Cruiser, Destroyer Gap In Upcming POMs," *Inside the Navy*, January 16, 2012. The POM is an internal DOD budget-planning document.

- congressional knowledge of the Flight III DDG-51 design; and

- DDG-51 design stability during the period covered by the proposed MYP arrangement.

Each of these factors is discussed below.

Potential for Additional Reductions to Planned Defense Spending Levels

Supporters of the proposed MYP arrangement could argue that, particularly given the potential under the Budget Control Act of 2011 (S. 365/P.L. 112-25 of August 2, 2011) for further reductions to planned levels of defense spending, policymakers should attempt to achieve savings in defense spending wherever possible, and that the savings that would be realized under the proposed MYP arrangement would contribute to this effort.

Skeptics of the proposed MYP arrangement could argue that given the potential for further reductions to planned levels of defense spending, it would be risky to enter into a commitment to procure a certain minimum number of DDG-51s over the next five years, and that using annual contracting, although more expensive than using an MYP arrangement, would give policymakers more flexibility for making changes in DDG-51 procurement rates in response to potential future reductions in defense spending.

Congressional Knowledge of Flight III Design

As mentioned earlier (see "Procurement of First Flight III DDG-51 Planned for FY2016" in "Background"), the Navy has announced that the Flight III design will be equipped with the AMDR. The exact size and capabilities of the AMDR, however, are not known, and little else has been announced about the configuration, capabilities, and cost of the Flight III design.

Supporters of the proposed MYP arrangement could argue that since the Flight III design will be a derivative of the well-known Flight IIA design, Congress already has significant knowledge about the configuration, capabilities, and cost of the Flight III design.

Skeptics of the proposed MYP arrangement could argue that the ultimate degree of difference between the Flight III design and the Flight IIA design is unclear—particularly given the uncertainty over the exact size and capabilities of the AMDR and the limited amount of other information available about the ship—and that it would be imprudent for Congress to commit now to the procurement in FY2016 and FY2017 of ships whose configuration, capabilities, and cost are not well understood.

Design Stability

The statute covering MYP contracts—10 U.S.C. 2306b—states that an MYP contract can be used for a DOD procurement program when the Secretary of Defense finds, among other things, "that there is a stable design for the property to be acquired and that the technical risks associated with such property are not excessive." For Navy shipbuilding programs, demonstrating the existence of a stable design traditionally has involved building at least one ship to that design and confirming, through that ship's construction, that the design does not need to be significantly changed. This is a principal reason why MYP contracts have not been used for procuring the lead ships in Navy shipbuilding programs.

Skeptics of using an MYP contract to procure the first Flight III DDG-51 could argue that the extent of design changes in the Flight III design—including the change in the ship's radar and associated changes in the ship's power-generation and cooling systems—make the ship different enough from the Flight IIA design that a stable design for the Flight III design has not yet been demonstrated. They can also note that the previous two DDG-51 MYP arrangements did not encompass a shift in DDG-51 flights, and that the first of these two MYP contracts began in FY1998, four years after the first Flight IIA DDG-51 was procured.

Supporters of using an MYP contract to procure the first Flight III DDG-51 could argue that notwithstanding design changes in the Flight III design, the vast majority of the DDG-51 design will remain unchanged, and that the stability of the basic DDG-51 design has been demonstrated through many years of production. They could also argue that although the two previous DDG-51 MYP contracts did not encompass a shift in DDG-51 flights, they nevertheless encompassed major upgrades in the design of the DDG-51.

Navy Perspective

The Navy's FY2013 budget submission states:

> The DDG 51 Class program is technically mature. To date 65 ships have been awarded, including 37 Flight IIA ships. Of the 65 ships awarded, 61 have delivered, and four are in construction. The program has successfully implemented capability upgrades during production while continuing to maintain configuration stability. The FY02-05 MYP ships included Baseline 7 Phase I.R combat system upgrade. The Baseline 7 Phase I.R combat system was introduced on the second FY02 ship (DDG 104). A total of 10 ships with the Baseline 7 Phase I.R combat system were awarded as part of the FY02-05 MYP. The FY98-FY01 MYP consisted of 13 ships. The SPY-1D radar on the 3rd ship of the MYP (DDG 91) was successfully replaced with the SPY-1D(V). This evolutionary approach allows the program to successfully incorporate the latest technologies while sustaining configuration stability and mitigating cost and schedule risk. At contract award, the nine ships proposed in this multiyear will be of the same configuration (Flight IIA). However, it is anticipated that one FY16 and two FY17 ships will incorporate Flight III capability as an engineering change proposal to mitigate the impact of MYP pricing. The Flight III ECP will not be awarded until the Flight III Milestone Decision Authority approves the configuration. The new Flight III radar (AMDR-S) will not be part of the multi-year procurement.

> The Flight III DDGs will utilize the same hull and major systems as current Flight IIA DDGs including LM 2500 propulsion gas turbines, Mk 41 Vertical Launch System, Mk 45 five inch Gun Weapon System, Mk 15 Phalanx Weapon System (CIWS), AN/SQQ-89 Undersea Warfare System and Tactical Tomahawk Weapon Control System. The principle dimensions and hull form will be unchanged from Flight IIA DDGs. The AN/SPY-1D(V) radar will be replaced with the AMDR-S radar and the ship's power and cooling systems will be upgraded to support the new radars. The deckhouse will be modified to accept the new radar arrays. The shipbuilding contracts will be fixed price incentive contracts, the same as previous DDG 51 Class ships. The overall ship design impact of these changes is estimated to be similar to those introduced on DDG 91 in FY98 as part of the FY98-FY01 MYP.[23]

---

[23] Department of the Navy, *Department of the Navy Fiscal year (FY) 2013 Budget Estimates, Justification of Estimates, Shipbuilding and Conversion, Navy*, February 2012, Exhibit MYP-1, Multiyear Procurement Criteria (MYP, Page 2 of 9), pdf page 148 of 246.

The Navy further stated on March 14, 2012, that

> The Baseline for the nine ships [included in the proposed MYP] will be established in the RFP in the same manner as prior DDG 51 MYP and annual procurements. The changes that constitute Flight III will not be part of that solicitation.
>
> The configuration changes covering Flight III and the Air and Missile Defense Radar (AMDR) will be introduced via Engineering Change Proposal (ECP) in a separate solicitation after the Flight IIA ship contracts have been awarded. While the timing of that RFP has not been finalized, it will not be released prior to the AMDR Milestone B decision scheduled for Fiscal Year 2013....
>
> In each of the two previous DDG 51 Class MYPs, significant Combat System changes were introduced as part of the ship solicitations. In the Fiscal Year 1998-2001 MYP, the AEGIS Weapon System (AWS) Baseline 7 Phase 1 was introduced on the third Fiscal Year 1998 ship, DDG 91. The upgrade included replacing the AN/SPY-1D radar with the AN/SPY-1D(V) radar, upgrading the Ship Service Gas Turbine Generators from 2500 KW to 3000 KW each, and the addition of a fifth 200 ton air conditioning plant. In the Fiscal Year 2002-2005 MYP, the AWS Baseline 7 Phase 1R was introduced on the second Fiscal Year 2002 ship, DDG 103. The upgrade replaced significant portions of the AWS computing architecture.
>
> The changes in the previous MYPs were provided to the shipbuilders in the form of ECPs and Design Budget Packages. These provided sufficient information to allow the shipbuilders to competitively bid the revised scope associated with the Combat System upgrades. The same process will be followed for the Fiscal Year 2013-2017 MYP.[24]

GAO Perspective

The January 2012 GAO report on DDG-51 acquisition stated:

> Despite uncertainty in the costs of the DDG 123, the Flight III lead ship, the Navy currently plans to buy the ship as part of a multiyear procurement, including 8 DDG Flight IIA ships, and award the contract in fiscal year 2013. Multiyear contracting is a special contracting method to acquire known requirements for up to 5 years if, among other things, a product's design is stable and technical risk is not excessive. According to the Navy, from fiscal year 1998 through 2005, the Navy procured Flight IIA ships using multiyear contracts yielding significant savings estimated at over $1 billion. However, the Navy first demonstrated production confidence through building 10 Flight IIAs before using a multiyear procurement approach. While Flight III is not a new clean sheet design, the technical risks associated with AMDR and the challenging ship redesign as well as a new power and cooling architecture coupled with the challenges to construct such a dense ship, will make technical risk high. Further, technical studies about Flight III and the equipment it will carry are still underway, and key decisions about the ship have not yet been made. DDG 123 is not due to start construction until fiscal year 2016. If the Navy proceeds with this plan it would ultimately be awarding a multiyear contract including this ship next fiscal year, even though design work has not yet started and without sufficient knowledge about cost or any construction history on which to base its costs, while waiting until this work is done could result in a more realistic understanding of costs. Our prior work has shown that construction of lead ships is

---

[24] Source: Navy information paper dated March 14, 2012, provided by the Navy to CRS on March 15, 2012.

challenging, the risk of cost growth is high, and having sufficient construction knowledge is important before awarding shipbuilding contracts....

If the Navy pursues a multiyear shipbuilding contract that includes the lead ship of Flight III, visibility over the risks inherent in lead ship construction could be obscured....

We also recommend that the Secretary of Defense take the following two actions:...

2. Ensure that the planned DDG 51 multiyear procurement request does not include a Flight III ship....

Regarding our fifth recommendation [number 2 above] that DOD not include a Flight III ship in its planned DDG 51 multiyear procurement request, DOD partially concurred, stating that it is following the statutory requirements for multiyear procurement authority. DOD commented that it will select an acquisition approach that provides flexibility and minimizes the cost and technical risk across all DDG 51 class ships. DOD expects to make a determination on including or excluding Flight III ships within the certification of the planned multiyear procurement that is due to Congress by March 1, 2012. While the Secretary can certify that due to exceptional circumstances, proceeding with a multiyear contract is in the "best interest" of DOD, notwithstanding the fact that one or more of the conditions of the required statutory certification are not met, requesting a multiyear procurement in March 2012 that includes the lead Flight III ship carries significant risk. DOD will be committing to a cost with no actual construction performance data on which to base its estimates and a ship concept and design that are not finalized. While DOD argued that it has in the past included DDG 51's that were receiving major upgrades in multiyear procurements, as this report shows, planned changes for Flight III could far exceed those completed in past DDG 51 upgrades. We therefore believe that, in view of the current uncertainty and risk, our recommendation remains valid to exclude a Flight III ship from the upcoming multiyear procurement request....

In the coming years, the Navy will ask Congress to approve funding requests for DDG 51 Flight III ships and beyond. Without a solid basis of analysis, we believe Congress will not have assurance that the Navy is pursuing an appropriate strategy with regard to its future surface combatants, including the appropriate level of oversight given its significant cost. To help ensure that the department makes a sound investment moving forward, Congress should consider directing the Secretary of Defense to:...

3. include the lead DDG 51 Flight III ship in a multi-year procurement request only when the Navy has adequate knowledge about ship design, cost, and risk.[25]

Adding a 10ᵗʰ Ship to the DDG-51 Multiyear Procurement (MYP)

Another issue for Congress concerns the possibility of adding a 10ᵗʰ DDG-51 to the proposed FY2013-FY2017 MYP arrangement for the DDG-51 program. Regarding this possibility, Sean Stackley, the Assistant Secretary of the Navy for Research, Development, and Acquisition (i.e., the Navy's acquisition executive), stated the following at a March 29, 2012, hearing on Navy shipbuilding programs before the Seapower and Projection Forces subcommittee, in response to a question about the FY2013 budget's deferral to FY2016 of a second DDG-51 that was previously programmed for FY2014:

[25] Government Accountability Office, *Arleigh Burke Destroyers[:] Additional Analysis and Oversight Required to Support the Navy's Future Surface Combatant Plans*, GAO-12-113, January 2012, pp. 48-50, 52-53, 54-55.

I'd like to address the question regarding the second destroyer in 2014. A couple of important facts: First, the—we restarted DDG-51 construction in—in [FY]2010 and we've got four ships under contract, and a result of the four ships that we've placed under contract is we have prior year savings in this program that are—work in our favor when we consider future procurement for the [DDG-]51s.

We also have a unique situation where we've got competition on this program—two builders building the 51s, and the competition has been healthy with both builders. We also have a very significant cost associated with government-furnished equipment, so not only did we restart construction at the shipyards, we also restarted manufacturing lines at our weapon systems providers.

So in this process we were able to restart 51s virtually without skipping a beat, and we're seeing the continued learning curve that we left off on back with the [FY]2005 procurement. So when we march into this third multiyear for the 51s we're looking to capitalize on the same types of savings that we saw prior, and our top line, again, allowed for nine ships to be budgeted, but when we go out with this procurement we're going to go out with a procurement that enables the procurement of 10 ships, where that 10th ship would be the second—potentially the second ship in [FY]2014 if we're able to achieve the savings that we're targeting across this multiyear between the shipbuilders in competition as well as the combat systems providers as well as all of the other support and engineering associated with this program.

So we want to leverage the strong learning, we want to leverage the strong industrial base, we want to leverage the competition to get to what we need in terms of both affordability and force structure, and I think we have a pretty good shot at it.[26]

Flight III DDG-51: Analytical Basis

Another issue for Congress is whether there is an adequate analytical basis for procuring Flight III DDG-51s in lieu of CG(X)s, and whether an analysis of alternatives (AOA) or the equivalent of an AOA should be performed before committing to the development and procurement of Flight III DDG-51s.[27]

---

[26] Source: Transcript of hearing. See also Megan Eckstein, "Navy Looking Into Feasibility Of Procuring 10th DDG In Multiyear Contract," *Inside the Navy*, April 2, 2012.

[27] The issue of whether there is an adequate analytical basis for canceling the CG(X) and instead procuring Flight III DDG-51s is somewhat similar to an issue raised by CRS several years ago as to whether there was an adequate analytical basis for the Navy's decision that a ship like the LCS—a small, fast ship with modular payload packages—would be the best or most cost-effective way to fill gaps the Navy had identified in its capabilities for countering submarines, small surface attack craft, and mines in heavily contested littoral areas. (See, for example, the September 5, 2002, update of CRS Report RS21305, *Navy Littoral Combat Ship (LCS): Background and Issues for Congress*, by Ronald O'Rourke, or the October 28, 2004, and the October 28, 2004, update of CRS Report RL32109, *Navy DDG-51 and DDG-1000 Destroyer Programs: Background and Issues for Congress*, by Ronald O'Rourke.)

The Navy eventually acknowledged that, on the question of what would be the best approach to fill these capability gaps, "the more rigorous analysis occurred after the decision to move to LCS." (Spoken testimony of Vice Admiral John Nathman, Deputy Chief of Naval Operations (Warfare Requirements and Programs), at an April 3, 2003, hearing on Navy programs before the Projection Forces subcommittee of the House Armed Services Committee. At this hearing, the chairman of the subcommittee, Representative Roscoe Bartlett, asked the Navy witnesses about the Navy's analytical basis for the LCS program. The witnesses defended the analytical basis of the LCS program but acknowledged that "The more rigorous analysis occurred after the decision to move to LCS." (See U.S. Congress, House Committee on Armed Services, Subcommittee on Projection Forces, *Hearing on National Defense Authorization Act for Fiscal Year 2004—H.R. 1588, and Oversight of Previously Authorized Programs*. 108th Cong., 1st (continued...)

Those who believe there is an adequate analytical basis for canceling the CG(X) and instead procuring Flight III DDG-51s could argue the following:

- Shifting to procurement of Flight III DDG-51s in FY2016, like shifting to procurement of Flight IIA DDG-51s in FY1994, would simply extend the DDG-51 production effort, and therefore would not amount to the initiation of a new shipbuilding program that would require an AOA or the equivalent of an AOA.

- The Navy's proposal to cancel the CG(X) and instead procure Flight III DDG-51s reflects substantial analytical work in the form of the CG(X) AOA, additional Navy studies that were done to support the 2008-2009 proposal to end DDG-1000 procurement and restart DDG-51 procurement, and the 2009 Navy destroyer hull/radar study that examined options for improving the AAW and BMD capabilities of the DDG-51 and DDG-1000 destroyer designs through the installation of an improved radar and combat system modifications.

Those who question whether there is an adequate analytical basis for canceling the CG(X) and instead procuring Flight III DDG-51s could argue the following:

- Procuring Flight III DDG-51s starting in FY2016 represents a significant change from the previous plan to procure CG(X)s starting around FY2017. Given the scope of the design modifications incorporated into the Flight III DDG-51 and the number of years that the design would be procured, the Navy's plan amounts to the equivalent of a new shipbuilding program whose initiation would require an AOA or the equivalent of an AOA.

- The CG(X) AOA focused mainly on examining radar and hull-design options for a cruiser with a large and powerful version of the AMDR, rather than radar- and hull-design options for a smaller destroyer with a smaller and less powerful version of the AMDR. The Navy's 2009 destroyer hull/radar study was focused on answering a somewhat narrowly defined question: what would be the lowest-cost option for improving the AAW and BMD performance of a DDG-51 or DDG-1000 by a certain amount through the installation of an improved radar and an associated modified combat system? An adequate analytical basis for a proposed program change of this magnitude would require an AOA or equivalent study that rigorously examined a broader question: given projected Navy roles and missions, and projected Navy and DOD capabilities to be provided by other programs, what characteristics of all kinds (not just AAW and BMD capability) are needed in surface combatants in coming years, and what is the most cost-effective acquisition strategy to provide such ships?

The January 2012 GAO report on DDG-51 acquisition stated:

> The Navy relied on its 2009 Radar/Hull Study as the basis to select DDG 51 over DDG 1000 to carry the Air and Missile Defense Radar (AMDR) as its preferred future surface combatant—a decision that may result in a procurement of up to 43 destroyers and cost up to

---

(...continued)

sess., March 27, and April 3, 2003, (Washington: GPO, 2003), p. 126. For an article discussing the exchange, see Jason Ma, "Admiral: Most LCS Requirement Analysis Done After Decision To Build," *Inside the Navy*, April 14, 2003.)

$80 billion over the next several decades. The Radar/Hull Study may not provide a sufficient analytical basis for a decision of this magnitude. Specifically, the Radar/Hull Study:

• focuses on the capability of the radars it evaluated, but does not fully evaluate the capabilities of different shipboard combat systems and ship options under consideration,

• does not include a thorough trade-off analysis that would compare the relative costs and benefits of different solutions under consideration or provide robust insight into all cost alternatives, and

• assumes a significantly reduced threat environment from other Navy analyses, which allowed radar performance to seem more effective than it may actually be against more sophisticated threats....

This study played a central role in determining future Navy surface combatant acquisitions by contributing to a selection of the Navy's preferred radar, combat system and ship solutions, making it, in essence, an AOA. Namely, the Radar/Hull Study provided analysis of the capability of multiple ship and radar alternatives against a revised IAMD capabilities gap, informing the selection of DDG 51 with AMDR as its preferred ship and radar combination. However, it does not provide an adequate evaluation of combat system and ship characteristics, and does not include key elements that are expected in an AOA that would help support a sound, long-term acquisition program decision.

Navy officials who were involved in the Radar/Hull Study told us that the capability of the technology concepts they evaluated was considered a major priority, and that the goal was identifying the most capable solution to meet the IAMD threat in the near-term that was also cost-effective. Within this context, the study team analyzed the capability of the radar variants considered. The Navy determined that a dual-band radar (S- and X-Band radars working together as an integrated unit) was required to effectively perform IAMD. As a result, the study team focused on assessing several different combinations of S- and X-Band radars....

The maximum radar size studied in the Radar/Hull Study was a 14-foot radar, since this was determined to be the largest size of radar that the DDG 51 hull could carry and the largest radar that DDG 1000 could carry without substantial deckhouse modifications. These radars were evaluated first against each other, and then combinations of radars were evaluated and compared with the capability of the current S-Band SPY-1D(V) radar installed on recent DDG 51 ships. All provided enhanced power over and above that of SPY-1D(V); this difference was quantified as a "SPY+" (in decibels) equating to the increase in target tracking range for a fixed amount of resources over the SPY-1D(V) radar. SPY+15 has a 32 times better signal to noise factor—or intensity of the returning radar signal echoing off a target over the intensity of background noise—than a SPY-1D(V) radar. Radars with additional average power and larger antennas have enhanced sensitivity, and thus better performance in advanced threat environments. The Navy found that the SPY+15 S-Band radars performed better than the SPY+11 S-Band radars, and the Radar/Hull Study's independent red team described the capability of SPY+15 as marginally adequate. The Navy also found that the AMDR-S performed IAMD better than the VSR+. For the X-Band, the Radar/Hull Study identified that SPY-3 performed better than SPQ-9B.

Although the Navy considered capability as a driving factor in its decision making, the Radar/Hull Study did not include a thorough comparative analysis of the capabilities of the two combat system architectures—Aegis on DDG 51 and the Total Ship Computing Environment (TSCE) on DDG 1000—into which the radars would need to be integrated. Other than assessing the BMD capability that Aegis currently possesses and the absence of BMD capability in TSCE, the Navy evaluated Aegis and TSCE by focusing on the amount of new software code that it estimated would be required to integrate the radars and to effectively perform IAMD and the costs and risks involved in this development. Such analysis is important because selection of a combat system essentially determines the ship choice, and the combat system is the interface between the radar and the ship's weapons. Since TSCE does not currently have an inherent BMD capability, the Navy identified several ways to add this capability using Aegis software and hardware. Similarly, changes were assessed to Aegis to provide it enhanced IAMD capability and the ability to leverage a dual-band radar....

Though TSCE was intended to be the combat system architecture for CG(X) and thus would have been modified to perform BMD, the Radar/Hull Study states that developing a BMD capability "from scratch" for TSCE was not considered viable enough by the study team to warrant further analysis, particularly because of the investment already made in the Aegis program. The Navy concluded that developing IAMD software and hardware specifically for TSCE would be more expensive and present higher risk. Ultimately, the Navy determined that Aegis was its preferred combat system option. Navy officials stated that Aegis had proven some BMD capability and was widely used across the fleet, and that the Navy wanted to leverage the investments it had made over the years in this combat system, especially in its current development of a version that provides a new, limited IAMD capability.

While the Navy's stated goal for the Radar/Hull Study was to identify the most capable solutions with an additional goal of affordability, the Navy selected Aegis based largely on its assessment of existing BMD capability, development costs and risk, and not on an analysis of other elements of combat system capability. Specifically, beyond the fact that Aegis already has a level of proven BMD capability and TSCE does not, other characteristics of the two combat systems that can contribute to overall performance were not evaluated.... Since this analysis was not conducted, any impact of these capabilities on IAMD or other missions or how each system compares with each other is unknown.

While considering the resident BMD capabilities of Aegis and comparing software development costs and risks are essential to making a decision, without a thorough combat system assessment, the Navy cannot be sure how other combat system characteristics can contribute to overall performance.

Because Aegis is carried by DDG 51 and not DDG 1000 ships, selection of Aegis as the preferred combat system essentially determined the preferred hull form. The Radar/Hull Study did not include any significant analysis of the ships themselves beyond comparing the costs to modify the ships to carry the new radar configurations and to procure variants of both types. Several characteristics associated with the ships (such as displacement or available power and cooling) were identified in the study.

The ships were evaluated on their ability to meet Navy needs and the impact of these ship characteristics on costs. However, there was no documented comparison or discussion of the benefits or drawbacks associated with any additional capabilities that either ship may bring. Navy officials told us that these characteristics were not weighted or evaluated against one another. Other ship variables that directly relate to ship capability and performance—such as damage tolerance and stealth features that were explicitly designed into DDG 1000—were not discussed in the Radar/Hull Study, even though they were discussed in the MAMDJF AOA. The MAMDJF AOA notes that a stealthy ship is harder for enemy forces to detect and target, thus making it more likely that a stealthy ship would be available to execute its BMD mission. However, senior Navy officials told us that the Radar/Hull Study did not consider the impact of stealth on performance because the study assumed that stealth would not have a significant impact on performance in IAMD scenarios. Navy officials added that any additional benefits provided by DDG 1000 stealth features were not worth the high costs, and that adding larger radars to DDG 1000 would reduce its stealth. However, no modeling or simulation results or analysis were presented to support this conclusion....

These characteristics [characteristics that were evaluated in the MAMDJF AOA that could have been evaluated in the Radar/Hull Study] influence performance, and each ship option has strengths and weaknesses that could have been compared to help provide a reasonable basis for selecting a ship. For example, DDG 1000 has enhanced damage survivability and reduced ship signatures, while DDG 51 is capable of longer time-on-station and endurance.

The Radar/Hull Study did not include a robust trade-off analysis for the variants studied to support the Navy's DDG 51 selection decision, which is currently planned to result in an acquisition of 22 modified Flight III DDG 51s and a further 21 modified DDG 51s known as Flight IV. DOD acquisition guidance indicates that a discussion of trade-offs between the attributes of each variant being considered is important in an AOA to support the rationale and cost-effectiveness of acquisition programs. A trade-off analysis usually entails evaluating the impact on cost of increasing the capability desired, essentially answering the question of how much more will it cost to get a greater degree of capability. A trade-off analysis allows decision makers to determine which combination of variables provides the optimal solution for a cost they are willing to pay. For the Radar/Hull Study, the Navy examined 16 different combinations of ship, radar, and combat system options based around DDG 51 and DDG 1000....

The Radar/Hull Study documents full cost data for only 4 of the 16 ship variants; 8 ship variants have no cost data, and 4 others do not have ship procurement and operations and support costs. Instead, the Radar/Hull Study provided full cost data for only the most expensive and least expensive DDG 51 and DDG 1000 variants (high and low), and operations and support costs for these four variants. Higher costs were largely driven by the combat system selected. For example, the high DDG 1000 variant included a 14-foot AMDR coupled with a SPY-3 radar, and the more expensive combat system solution, which comprised replacing the central core of DDG 1000's TSCE combat system with the core of the Aegis combat system. The high DDG 51 variant included a 14-foot AMDR coupled with a SPY-3 radar and the Aegis combat system. The low DDG 1000 variant coupled a 12-foot VSR+ with the SPY-3 radar and a less expensive combat system solution involving replacing only portions of TSCE with portions of Aegis. The low DDG 51 included VSR+ coupled with the SPQ-9B radar and the Aegis combat system. In both the DDG 1000 high and low cases, the combat system solutions would be equally capable; the difference was in the level of effort and costs required to implement the changes. Since only a high and low version of DDG 1000s were priced out, the study did not include a DDG 1000 variant with AMDR and the less complicated TSCE combat system upgrade that may be a less expensive—but equally capable—option. Because this variant was not included in the study, cost data were not provided. This study also presented a brief analysis of operations and support costs; the Navy concluded that it found only negligible differences between the operations and support costs for the DDG 51 and DDG 1000 variants. Previous DDG 1000 cost estimates had indicated 28 percent lower long-term costs than DDG 51. While both ships had increases in these costs, the Navy determined in the Radar/Hull Study that adding additional crew to DDG 1000 to perform BMD-related tasks and increased fuel costs were more significant for that ship, and made the costs essentially equal between the two ships....

Navy officials agreed that they could have developed cost estimates for all 16 of the variants, but stated that there was a time constraint for the study that prohibited further analysis, and that they believed that pricing the high and low options was enough to bound the overall costs for each ship class. Without complete cost data for all variants, the Navy could not conduct a thorough trade-off analysis of the variants that fell between the high and low extremes because the costs of these variants are unknown. DOD acquisition guidance highlights the importance of conducting a trade-off analysis. Conducting a trade-off analysis with costs for all the variants would have established the breakpoints between choices, and identified potential situations where a cheaper, slightly less capable ship or a more expensive but much more capable ship might be a reasonable choice....

Further, the Navy also did not prioritize what aspects of the radar, combat system, and ship it valued more than others, which could also be used to inform a trade-off analysis. For example, if performance is valued more than cost, choosing a ship variant that has 10 percent more performance than another variant but with a 20 percent increase in cost might be in the Navy's best interest. Alternatively, if cost was weighted more than performance, the Navy might choose the cheaper and slightly less capable ship as it would be able to get a 20 percent reduction in cost with only a 10 percent reduction in performance. Similarly, the study did not discuss the Navy's preferences with regard to ship characteristics and the impact that differences in these characteristics might have on a trade-off analysis. For example, Navy officials told us that electrical power was a major concern for future destroyers, but the considerable difference in available power between DDG 51 and DDG 1000 (approximately 8,700 kilowatts for DDG 51 after the addition of a supplemental generator required) was not compared in a trade-off analysis. Finally, the Navy did not assess potential impacts of ship selection on future fleet composition. The MAMDJF AOA found that more capability can be obtained by fewer, more capable ships (meaning those with larger radars) than a greater number of less capable ships (meaning those with smaller radars). This could change the acquisition approach and would result in different program costs as a result if it is found that fewer, more capable ships are more cost-effective than many, less capable ships.

Navy officials told us that some of these trade offs were not done in the Radar/Hull Study because they were already studied in the MAMDJF AOA. However, that study, using a different threat environment and ship concepts, eliminated the DDG 51 variant from further consideration as a single ship solution; it also eliminated the DDG 1000 option without a radar larger than the 14-foot design that was considered in the Radar/Hull Study. Consequently, its analysis is not directly comparable or interchangeable with the Radar/Hull Study. When comparing the raw ship data from the Radar/Hull Study, we found that the two ships offer different features worth evaluating. For example, all DDG 1000 variants offer more excess cooling and service life allowance, meaning the ability of the ship to accommodate new technologies over the life of the ship without major, costly overhauls than DDG 51 variants, while DDG 51 variants offer greater endurance and lower procurement costs....

As this table shows, these two ships offer different characteristics. Both were deemed capable of carrying AMDR, but without conducting a trade-off analysis of these characteristics, the Navy did not consider their relative merit and the significance, if any, of any differences between the two. Senior Navy officials told us that it is now conducting these types of trade-off analyses; however, these analyses are focused only on assessing various DDG 51 configurations, and were not done to help inform the ship selection decision. A preliminary finding of these new analyses is that the cost of Flight III is estimated to range from $58 billion to $64 billion (in constant 2012 dollars), including research and development and procurement.

The Radar/Hull Study assumed a significantly reduced threat environment compared to the earlier MAMDJF AOA and other Navy studies. How the threat is characterized is important because against a reduced threat environment, a less capable radar than what was identified as necessary in the MAMDJF AOA was described by the Radar/Hull Study as marginally adequate. Both the Radar/Hull Study and MAMDJF AOA analyzed the performance of radars in several different classified tactical situations that presented threats of varying levels of complexity. The most stressing situations involved a number of different air and missile threats and a complex timing of events. In the MAMDJF AOA, these tactical situations involved many different types of simultaneous threats and larger radars, and were developed in consultation with the Office of Naval Intelligence—the agency tasked to provide validated threat intelligence to support Navy and joint, Navy-led acquisition programs—as well as MDA. Conversely, the subsequent Radar/Hull Study assumed a significantly reduced threat environment and smaller radar solutions than did the MAMDJF AOA. This study modeled radar performance based on a very limited air and missile threat which are both quantitatively and qualitatively less stressing than the threat environment established in the MAMDJF AOA, in other Navy and DOD threat analyses, and in system guideline documents for AMDR. Also, the Office of Naval Intelligence was not actively engaged in the Radar/Hull Study. The system guideline documents for AMDR that were generated at approximately the same time as the Radar/Hull Study also included significantly more taxing tactical situations than the Radar/Hull Study, and in some cases they are even more stressing than those found in the MAMDJF AOA. The Office of Naval Intelligence also provided input to these AMDR system guidelines.

The Navy believes that some of the differences in the threat environment result from the different timeframes for the Radar/Hull Study and the MAMDJF AOA; the MAMDJF AOA states that it is based on a 2024 through 2030 timeframe while the Radar/Hull Study states that it is based on a 2015 through 2020 timeframe. However, Navy officials also told us that the IAMD threats are actually emerging more rapidly than they had assumed in the MAMDJF AOA, which could mean that some of the MAMDJF AOA threats may be present earlier. The Navy does not document why the Radar/Hull Study based its analysis on a reduced threat environment compared to the MAMDJF AOA, since both studies are attempting to identify solutions to the same capabilities gap and set of requirements. Navy officials later told us that the assumption in the Radar/Hull Study was that no single Navy ship would likely have to deal with all the threats in the battlespace, compared to the threat environment in the MAMDJF AOA where more of a single-ship solution was considered. However, other Navy studies developed in a similar timeframe to the Radar/Hull Study describe a larger number of threats than the Radar/Hull Study. Further, while the Navy's assumption may account for some of the quantitative differences between the Radar/Hull Study and all the other Navy studies we analyzed, it should have no bearing on the qualitative difference in the composition of the threat, since this is a variable that is independent of Navy concepts of operations and is a variable over which the Navy has no influence....

The Navy is in the early stages of a potential $80 billion investment in up to 43 DDG 51 destroyers to provide IAMD capability for potentially up to the next 60 years. Such investment decisions cannot be made without some degree of uncertainty; they will always involve risks—especially in the early stages of a program. Yet, a decision of this magnitude should proceed with a solid base of analysis—regarding the alternatives, cost, and technical risks—as well as a plan for oversight that provides sufficient leverage and flexibility to adapt to information as it emerges. These pieces are not sufficiently in place, at least with respect to Flight III and AMDR. To its credit, the Navy's goal was to move towards a lower-cost solution that could be rapidly fielded; however, there are a number of key shortfalls in the Navy's analysis in support of its decisions. As it stands, the Navy risks getting a solution that is not low cost, will not be fielded in the near-term, or meet its long-term goals. DDG 51 may ultimately be the right decision, but at this point, the Navy's analysis has not shown this to be the case. Specific issues include:

• The Navy's choices for Flight III will likely be unsuitable for the most stressful threat environments it expects to face.

• While the Navy potentially pursued a lower-cost ship solution, it did not assess the effect of this decision in terms of long-term fleet needs where more of these ships may be required to provide the same capability of a smaller number of more costly, but more capable, ships.

• Though the Navy hopes to leverage sensor netting to augment the capability of these ships, there is a shortage of analysis and testing with operational assets to demonstrate that this is a viable option.

• The Navy clearly states in recent AMDR documents that a new, as-of-yet undefined ship is required to meet its desired IAMD capability. However, it has not yet articulated its long-term plans for a new surface combatant that is sized to be able to carry a larger AMDR, and such a ship is not currently in the Navy's long-range shipbuilding plan.

• Without a robust operational test program that will demonstrate both DDG 51 with the modified Aegis combat system and the new AMDR, the Navy cannot be sure that the ships can perform the IAMD mission as well as planned....

We recommend that the Secretary of Defense direct the Secretary of the Navy to take the following three actions:

1. Conduct a thorough AOA in accordance with DOD acquisition guidance for its future surface combatant program to include: (a) a range of representative threat environments developed in concert with the intelligence community; (b) results of its ongoing Flight III studies and full cost estimates in advance of awarding DDG 51 Flight III production contracts; (c) implications of the ability of the preferred ship to accommodate new technologies on future capabilities to determine the most suitable ship to carry AMDR and meet near-term IAMD requirements and provide a path to far-term capabilities; (d) implications on future fleet composition; and (e) an assessment of sensor netting—conducted in consultation with MDA and other cognizant DOD components—to determine the risks inherent in the sensor netting concept, potential current or planned programs that could be leveraged, and how sensor netting could realistically be integrated with the selected future surface combatant to assist in conducting BMD. This AOA should be briefed to the Joint Requirements Oversight Council....

DOD did not agree with our first recommendation to conduct an AOA to support its future surface combatant selection decision, stating that its previous analyses—specifically the MAMDJF AOA and the Radar/Hull Study—comprise a body of work that satisfies the objectives of an AOA. However, DOD did not present any additional evidence to refute our findings. DOD did agree that an assessment of sensor netting needs to be performed. Our analysis shows that the Radar/Hull Study, which was the key determinant in the DDG 51 decision, was a departure from the MAMDJF AOA. These studies are neither complementary nor can they be aggregated. While both sought to determine the best solution to address identified integrated air and missile defense gaps, the Radar/Hull Study essentially answered a different question than the MAMDJF AOA. In essence, it was attempting to identify a cost-constrained, less robust solution, which makes analysis from one study not always appropriate to apply to the other. Specifically, the MAMDJF AOA considered a significantly more taxing threat environment than the Radar/Hull Study, requiring ships carrying very large radars to independently manage these threats. Alternatively, the Radar/Hull Study considered a much less taxing threat environment, allowing for ships carrying smaller radars but that would need to work together to be effective. Ultimately, the MAMDJF AOA eliminated DDG 51 from consideration as a single-ship solution. DOD also states that it is currently conducting additional studies on Flight III, but since these are solely focused on DDG 51, they do not provide any additional insight into the decision as to the appropriate ship that might be used to supplement the Navy's existing analysis. As we note in this report, the proposed program calls for an investment of up to approximately $80 billion for 43 destroyers, and likely more if the Navy chooses to pursue a Flight IV concept. Given the scope of the Navy's plans, a thorough AOA is essential to affirm that the decision made is the right one and a sound investment moving forward. This AOA should be briefed to Joint Requirements Oversight Council because of the magnitude of this potential acquisition and because of the joint service interest in IAMD that make it important to have an overarching body review the Navy's analysis and decisions. We believe that this recommendation remains valid....

In the coming years, the Navy will ask Congress to approve funding requests for DDG 51 Flight III ships and beyond. Without a solid basis of analysis, we believe Congress will not have assurance that the Navy is pursuing an appropriate strategy with regard to its future surface combatants, including the appropriate level of oversight given its significant cost. To help ensure that the department makes a sound investment moving forward, Congress should consider directing the Secretary of Defense to:

1. require the Navy to submit a thorough, well-documented AOA for the its future surface combatant program that follows both DOD acquisition guidance and the elements outlined in our first recommendation prior to issuing solicitations for any detail design and construction contracts of DDG 51 Flight III ships....[28]

Flight III DDG-51: Adequacy of AAW and BMD Capability

Another issue for Congress is whether the Flight III DDG-51 would have sufficient AAW and BMD capability to adequately perform future AAW and BMD missions.

The Flight III DDG-51 would have more AAW and BMD capability than the current DDG-51 design, but less AAW and BMD capability than was envisioned for the CG(X), in large part because the Flight III DDG-51 would be equipped with a 12- or 14-foot-diameter version of the AMDR that would have more sensitivity than the SPY-1 radar on Flight IIA DDG-51s, but less sensitivity than the substantially larger version of the AMDR that was envisioned for the CG(X). The CG(X) also may have had more missile-launch tubes than the Flight III DDG-51.

Supporters of the Navy's proposal to procure Flight III DDG-51s could argue that a 12- or 14-foot-diameter version of the AMDR would provide the DDG-51 with sufficient AAW and BMD capability to perform projected AAW and BMD missions because this radar would be substantially more capable than the SPY-1 radar currently on DDG-51s, and because Flight III DDG-51s (and other Navy ships) would also benefit from data collected by other sensors, including space-based sensors.

Skeptics could argue that Flight III DDG-51s might not have sufficient AAW and BMD capability because a 12- or 14-foot-diameter AMDR would be substantially less capable than the substantially larger AMDR that the Navy previously believed would be needed to adequately perform projected AAW and BMD missions, because the off-board sensors on which the Flight III DDG-51 would rely for part of its sensor data that might turn out to be less capable as the Navy assumed in 2008 that they would be, and because the off-board sensors and their related data-communication links could in any event be vulnerable to enemy attack.

The January 2012 GAO report on DDG-51 acquisition stated that

---

[28] Government Accountability Office, *Arleigh Burke Destroyers[:] Additional Analysis and Oversight Required to Support the Navy's Future Surface Combatant Plans*, GAO-12-113, January 2012, summary page and pp. 7-19, 51, 53-54, 55.

the Navy's choice of DDG 51 as the platform for AMDR limits the overall size of the radar to one that will be unable to meet the Navy's desired (objective) IAMD [integrated air and missile defense] capabilities. If the Navy selects a 12-foot AMDR—which may reduce the impacts on the ship and design—it may not be able to meet the requirements for AMDR as currently stated in the Navy's draft capabilities document....

Flight III with a 14-foot AMDR will not be powerful enough to meet the Navy's objective, or desired IAMD capabilities. The shipyards and the Navy have determined that 14-foot radar arrays are the largest that can be accommodated within the confines of the existing DDG 51 configuration. Adding a radar larger than 14 feet to DDG 51 is unlikely without major structural changes to the ship. AMDR is being specifically developed to be a scalable radar—meaning that it can be increased in size and power to provide enhanced capability against emerging threats.

According to AMDR contractors, the Navy had originally contracted for an investigation of a Variant 2 AMDR with a sensitivity of SPY+40, but this effort was cancelled. They added that the maximum feasible size of AMDR would be dictated by the ship and radar power and cooling demands, but that they had investigated versions as large as 36 feet. Leveraging AMDR's scalability will not be possible on DDG 51 without major changes, such as a new deckhouse or adding to the dimensions of the hullform itself by broadening the beam of the ship or adding a new section (called a plug) to the middle of the ship to add length. Navy officials have stated that adding a plug to DDG 51 is not currently a viable option due to the complexity, and that a new ship design is preferable to a plugged DDG 51.

The Navy has not yet determined the size of AMDR for Flight III, and two sizes are under consideration: a 14-foot AMDR with a sensitivity of SPY+15, and a 12-foot AMDR with a sensitivity of SPY+11. According to a draft AMDR Capability Development Document, the Navy has identified that an AMDR with SPY+15 will meet operational performance requirements against the threat environment illustrated in the Radar/Hull Study. This document also notes that a significantly larger SPY+30 AMDR is required to meet the Navy's desired capability (known as objective) against the threat environment illustrated in the MAMDJF AOA. The Navy could choose to change these requirements. The MAMDJF AOA eliminated the DDG 51-based SPY+15 solution from consideration in part due to the limited radar capability, and identified that a radar closer to SPY+30 power with a signal to noise ratio 1,000 times better than SPY+0 and an array size over 20 feet is required to address the most challenging threats. If a 12-foot array is chosen, the Navy will be selecting a capability that is less than the "marginally adequate" capability offered by a SPY+15 radar as defined by the Radar/Hull Study red team assessment. According to Navy officials, only through adding additional square footage can the Navy effectively make large improvements in the sensitivity of the radar the SPY+30 radar considered in the MAMDJF AOA could only be carried by a newly designed cruiser or a modified San Antonio class ship, and only a modified DDG 1000 and could carry the approximately SPY+25 radar. According to the draft AMDR Capability Development Document, the Navy's desired IAMD capability can only be accommodated on a larger, currently unspecified ship. As part of the MAMDJF AOA, the Navy identified that DDG 1000 can accommodate a SPY+25 radar. As part of a technical submission to the Navy, BIW—the lead designer for DDG 1000—also identified a possible design for a 21-foot radar on DDG 1000. The Navy did not include a variant with this size radar in the Radar/Hull Study.

According to senior Navy officials, since the MAMDJF AOA was released the Navy has changed its concept on the numbers of Navy ships that will be operating in an IAMD environment. Rather than one or a small number of ships conducting IAMD alone and independently managing the most taxing threat environments without support, the Navy now envisions multiple ships that they can operate in concert with different ground and space-based sensor assets to provide cueing for AMDR when targets are in the battlespace. This cueing would mean that the shooter ship could be told by the off-board sensors where to look for a target, allowing for earlier detection and increased size of the area that can be covered. According to the Navy, this concept—referred to as sensor netting—can be used to augment the reduced radar capability afforded by a 12 or 14-foot AMDR as compared to the larger radars studied in the MAMDJF AOA. For example, the Navy cited the use of the Precision Tracking Space System program as an example of sensors that could be leveraged. However, this program (envisioned as a constellation of missile tracking satellites) is currently in the conceptual phase, and the independent Radar/Hull Study red team stated that the development timeline for this system is too long to consider being able to leverage this system for Flight III. Navy officials told us that another option would be to leverage the newly completed Cobra Judy Replacement radar ship and its very powerful dual-band radar to provide cueing for DDG 51s. This cueing could allow the DDG 51s to operate a smaller AMDR and still be effective. The Cobra Judy Replacement ship is comparatively cheaper than DDG 51s (approximately $1.7 billion for the lead ship), and was commercially designed and built. However, it is not a combatant ship, which would limit its employment in a combat environment and make it difficult to deploy to multiple engagement locations.

Senior Navy officials told us that the concept of sensor netting is not yet well defined, and that additional analysis is required to determine what sensor capabilities currently exist or will be developed in the future, as well as how sensor netting might be conceptualized for Flight III. Sensor netting requires not only deployment of the appropriate sensors and for these sensors to work alone, but they also need to be able to share usable data in real-time with Aegis in the precise manner required to support BMD engagements. Though sharing data among multiple sensors can provide greater capabilities than just using individual stand-alone sensors, officials told us that every sensor system has varying limitations on its accuracy, and as more sensors are networked together and sharing data, these accuracy limitations can compound. Further, though there have been recent successes in sharing data during BMD testing, DOD weapons testers responsible for overseeing BMD testing told us that there have also been issues with sending data between sensors. Although sensor technology will undoubtedly evolve in the future, how sensor netting will be leveraged by Flight III and integrated with Navy tactics to augment Aegis and the radar capability of Flight III is unknown...

The Navy's choices for Flight III will likely be unsuitable for the most stressful threat environments it expects to face....

We recommend that the Secretary of Defense direct the Secretary of the Navy to take the following three actions:...

2. Report to Congress in its annual long-range shipbuilding plan on its plans for a future, larger surface combatant, carrying a more capable version of AMDR and the costs and quantities of this ship....

DOD concurred with our second recommendation that the Navy report to Congress in its annual long-range shipbuilding plan on its plans for a future larger surface combatant carrying a more capable version of AMDR. Given the assessments that the Navy is currently conducting on surface combatants, the Navy's next submission should include more specific information about its planned future surface combatant acquisitions.[29]

---

[29] Government Accountability Office, *Arleigh Burke Destroyers[:] Additional Analysis and Oversight Required to Support the Navy's Future Surface Combatant Plans*, GAO-12-113, January 2012, pp. 31, 41-44, 52, 53.

Flight III DDG-51: Cost, Technical, and Schedule Risk

Another issue for Congress concerns cost, technical, and schedule risk for the Flight III DDG-51. Some observers have expressed concern about the Navy's ability to complete development of the AMDR and deliver the first AMDR to the shipyard in time to support the construction schedule for a first Flight III DDG-51 procured in FY2016. The Navy could respond to a delay in the development of the AMDR by shifting the procurement of the first Flight III DDG-51 to FY2017 or a later year, while continuing to procure Flight IIA DDG-51s. (The MYP that the Navy is proposing for FY2013-FY2017 is structured to accommodate such a shift, should it become necessary.) Observers have also expressed concern about the potential procurement cost of the Flight III DDG-51 design.

July 2012 CBO report

A July 2012 Congressional Budget Office (CBO) report on the cost of the FY2013 Navy 30-year (FY2013-FY2042) shipbuilding plan states that

> a DDG-51 Flight III would cost $700 million, or about 40 percent, more than a new Flight IIA, by CBO's estimate. Thus, CBO estimates, the average cost per ship would be $2.4 billion....
>
> CBO's estimate of the cost of each of those ships is about $200 million more than it was last year. Most of the increase stemmed from updated information on the cost of incorporating the AMDR into the Flight IIA configuration. At the same time, the Navy decreased its estimate for the average price of a DDG-51 Flight III from $2.4 billion in the [FY]2012 [30-year shipbuilding] plan to $2.2 billion in the [FY]2013 [30-year shipbuilding] plan, primarily by incorporating the use of multiyear procurement authority in its estimates, as it did for all destroyers bought between 1998 and 2005. Considerable uncertainty remains in the DDG-51 Flight III program, however. Costs could be substantially higher or lower than CBO's estimate, depending on how well the restart of the DDG-51 program goes and on the eventual cost and complexity of the AMDR and associated changes in the ship's design.[30]

March 2012 Hearing

At a March 29, 2012, hearing on Navy shipbuilding programs before the Seapower and Projection Forces subcommittee of the House Armed Services Committee, Department of the Navy officials testified in written form that

> AMDR technology development is on track and successfully completed the three System Functional Reviews in December 2011. Prototype development to demonstrate critical technologies is well underway. The program remains on schedule for the Preliminary Design Reviews in the fall of 2012 and the Navy plans to award an Engineering and Manufacturing Development contract in early Fiscal Year 2013.[31]

---

[30] Congressional Budget Office, *An Analysis of the Navy's Fiscal Year 2013 Shipbuilding Plan*, July 2012, p. 22.

[31] Statement of the Honorable Sean J. Stackley, Assistant Secretary of the Navy (Research, Development and Acquisition) and Vice Admiral John Terrence Blake, Deputy Chief of Naval Operations For Integration of Capabilities and Resources and Lieutenant General Richard P. Mills, Deputy Commandant, Combat Development and Integration & Commanding General, Marine Corps Combat Development Command, before the Subcommittee on Seapower and Projection Forces of the House Armed Services Committee [hearing] on Navy Shipbuilding Acquisition Programs and (continued...)

In his spoken testimony at the hearing, one of the Department of the Navy witnesses—Sean Stackley, the Assistant Secretary of the Navy for Research, Development, and Acquisition (i.e., the Navy's acquisition executive)—stated the following when asked about the status of the AMDR development effort:

> The development is going great. We've got three industry competitors that are working on the development. They've each been able to leverage other systems that have been developed using the technology associated with the AMDR radar, and so when we kicked off the competition they were well out in front in terms of level of maturity of the technology.
>
> We're going through—I'll call it a small scale prototype development to demonstrate, you know, proficiency of the respective designs that will be leading to a downselect. I am very upbeat on the progress we're making on AMDR and I'm highly confident that that program is—is right on step to support introduction [of the AMDR] on [the FY]2016 DDG-51.[32]

Another witness at the hearing—Phebe N. Novakovic, executive vice president, marine systems for General Dynamics—stated the following in response to a question about the cost of the Flight III DDG-51 and the integration of the AMDR into the Flight III design:

> We are approaching this [DDG-51] multiyear [for FY2013-FY2017], should we get that authority, and certainly this block of ships in the same way that we have approached each competition that we're in. I don't see any particular additional uncertainty as we think about these ships.
>
> We do not understand that radar. We're going to have to understand it better and its—and its interfaces with the ship. We've got a long way to go until we're at that point where we need to go bid and size. That we work with the needy [sic: Navy] customer and, frankly, the—the other industry partners who have been very, very helpful.
>
> I'll give you an example. We're doing the combat systems integration with Raytheon. We have a tiger team with Raytheon because we're not electronics guys, right? So the extent— they've been very useful in teaching us a lot about how their systems work so we could optimize the integration of that system into the ship hull.
>
> It's that kind of process that we'll apply to a—whatever the—the changes are in the—and even if they're substantial changes—in the configuration of the [DDG-]51s. So we can bid as long as we understand we can—and understand that—the risks areas and that—that you're properly protected around those risk areas, and everybody's—everybody's reasonable about understanding what they are I think is ... [sic][33]

March 2012 GAO Report

A March 2012 GAO report assessing selected DOD acquisition programs stated:

> According to the Navy, all six critical technologies for the AMDR program are expected to be nearing maturity and demonstrated in a relevant environment before a decision is made to

---

(...continued)

Budget Requirements of the navy's Shipbuilding and Construction Plan, March 29, 2012, p. 7.

[32] Source: Transcript of hearing.

[33] Source: Transcript of hearing. The transcript shows the quoted statement ending with an ellipsis.

---

enter system development. They are currently immature. Program officials stated that digital beamforming technology—necessary for AMDR's simultaneous air and ballistic missile defense mission—has been identified as the most significant challenge and will likely take the longest time to mature. Digital beamforming enables the radar to generate and process multiple beams simultaneously, which results in more radar resources being available to support simultaneous air and missile defense. Program officials stated that this technology has been used before, but it has never been demonstrated in a radar as large as AMDR.

The AMDR's transmit-receive modules—the individual radiating elements key to transmitting and receiving electromagnetic signals—also pose a challenge. According to the program office, similar radar programs have experienced significant problems developing transmit-receive modules, resulting in cost and schedule growth. To achieve the increased performance levels required for AMDR, the contractor will likely use gallium nitride semiconductor technology instead of the legacy gallium arsenide technology. The new technology has the potential to provide higher power and efficiency with a smaller footprint. According to the Navy, this would reduce the power and cooling demands placed on the ship by the radar. However, gallium nitride has never been used in a radar as large as the AMDR, and long-term reliability and performance of this newer material is unknown. If these transmit-receive units cannot provide the required power, the program would either need to use the legacy technology and increase the power and cooling resources available for the AMDR, or accept reduced power and performance for the AMDR S-band radar.[34]

January 2012 GAO Report

The January 2012 GAO report on DDG-51 acquisition stated:

> The Navy plans to procure the first of 22 Flight III DDG 51s in 2016 with the new AMDR and plans to achieve Flight III initial operational capability in 2023. Other than AMDR, the Navy has not identified any other technologies for inclusion on Flight III or decided on the size of AMDR. Although the analysis supporting Flight III discusses a 14-foot AMDR, senior Navy officials recently told us that a 12-foot AMDR may also be under consideration. While the Navy is pursuing a thoughtful approach to AMDR development, it faces several significant technical challenges that may be difficult to overcome within the Navy's current schedule. The red team assessment of an ongoing Navy Flight III technical study found that the introduction of AMDR on DDG 51 leads to significant risks in the ship's design and a reduced future capacity and could result in design and construction delays and cost growth on the lead ship.... Given the level of complexity and the preliminary Navy cost estimates, the Navy has likely underestimated the cost of Flight III....

> AMDR represents a new type of radar for the Navy, which the Navy believes will bring a significantly higher degree of capability than is currently available to the fleet. AMDR is to enable a higher degree of IAMD than is possible with the current legacy radars. Further, the Navy believes that through the use of active electronically scanned array radars, AMDR will be able to "look" more places at one time, thus allowing it to identify more targets with better detection sensitivity.26

> • AMDR-S: a 4 faced S-Band radar providing volume search for air and ballistic missile defense; It will also allow the radar to view these targets with better resolution. AMDR is conceived to consist of three separate parts:

---

[34] Government Accountability Office, *Defense Acquisitions[:] Assessments of Selected Weapon Programs*, GAO-12-400SP, March 2012, p. 44.

• AMDR-X: a 3 faced, 4-foot by 6-foot X-Band radar providing horizon search (as well as other tasks such as periscope and floating mine detection); and

• Radar suite controller: interface to integrate the two radars and interface with the combat system....

Three contractors are under contract to mature and demonstrate the critical AMDR-S radar technology required; the acquisition of the AMDR-X portion is still in the preliminary stage, and the Navy plans to award a contract for it in fiscal year 2012.

The Navy recognized the risks inherent in the AMDR-S program early on, and implemented a risk mitigation approach to help develop and mature specific radar technologies that it has identified as being particularly difficult. Additionally, the Navy used an initial AMDR-S concept development phase to gain early contractor involvement in developing different concepts and earlier awareness of potential problems. In September 2010, the Navy awarded three fixed-priced incentive contracts to three contractors for a 2-year technology development phase. All three contractors are developing competing concepts with a goal of maturing and demonstrating S-Band and radar suite controller technology prototypes. In particular, the contractors are required to demonstrate performance and functionality of radar algorithms in a prototype one-fifth the size of the final AMDR-S.

The Navy has estimated that AMDR will cost $2.2 billion for research and development activities and $13.2 billion to procure at most 24 radars. At the end of the 2-year phase, the Navy will hold a competition leading to award of an engineering and manufacturing contract to one contractor....

AMDR is first scheduled to be delivered to a shipyard in fiscal year 2019 in support of DDG 123—the lead ship of Flight III....

AMDR-S relies on several cutting-edge technologies. Three of the most significant of these pertain to digital beamforming, the transmit/receive modules, and the radar/combat system interface....

Though the Navy has been pursuing risk mitigation efforts related to some key AMDR technologies, realizing AMDR will require overcoming several significant technical challenges. For example, though the Navy worked with the United Kingdom on a radar development program to demonstrate large radar digital beamforming, including limited live target testing, the technical challenges facing the development of AMDR have not been fully mitigated by these efforts. The joint radar development program used a digital beamforming architecture different than what is intended for AMDR, and the demonstrator was much smaller than what is envisioned for AMDR-S. Further, the Navy's previous effort also did not demonstrate against BMD targets, which are the most stressful for the radar resources. The Navy told us that the contractors have been successful in their AMDR development efforts to date, and that power and cooling requirements may be less than initially estimated. However, substantial work remains, and failure to achieve any of these technologies may result in AMDR being less effective than envisioned. AMDR development is scheduled for 10 years, compared with 9 years for the DDG 1000's VSR.

Integration with the Aegis combat system may also prove challenging: Aegis currently receives data from only a single band SPY-1D(V) radar, and adding AMDR will require modifying Aegis to receive these data, to accommodate some new capabilities, and to integrate Aegis with the radar suite controller. The Navy has deferred this integration, as it recently decided to eliminate AMDR integration work from its upcoming Aegis upgrade (ACB 16) contract, although Navy officials pointed out that this work could be started later under a separate contract. If the Navy does not fund AMDR integration work in ACB 16, this work may not be under way until the following ACB upgrade, which could be completed in 2020 at the earliest if the Navy remains on the same 4 year upgrade schedule. With an initial operating capability for Flight III planned for 2023, this could leave little margin for addressing any problems in enabling AMDR to communicate with the combat system.

DDG 51 is already the densest surface combatant class; density refers to the extent to which ships have equipment, piping, and other hardware tightly packed within the ship spaces According to a 2005 DOD-sponsored shipbuilding study, the DDG 51 design is about 50 percent more dense and complex than modern international destroyers. High-density ships have spaces that are more difficult to access; this results in added work for the shipbuilder since there is less available space to work efficiently. As a legacy design, the ship's physical dimensions are already fixed, and it will be challenging for the Navy to incorporate AMDRs' arrays and supporting equipment into this already dense hullform. Some deckhouse redesign will be necessary to add the additional radar arrays: a current DDG 51 only carries four SPY radar arrays, while Flight III is envisioned to carry four AMDR-S arrays plus three additional AMDR-X band arrays. The deckhouse will need to be redesigned to ensure that these arrays remain flush with the deckhouse structure. Adding a 14-foot AMDR to DDG 51 will also require significant additional power generating and cooling equipment to power and cool the radar. Navy data show that as a result of adding AMDR the ships will require 66 percent more power and 81 percent more cooling capacity than current DDG 51s. If the Navy elects to use a smaller AMDR for Flight III these impacts may be reduced, but the ship would also have a significant reduction in radar performance.

The addition of AMDR and the supporting power and cooling equipment will significantly impact the design of Flight III. For example, additional large cooling units—each approximately 8 feet by 6 feet—required to facilitate heat transfer between the radar coolant and the ship's chilled water system will have to be fit into the design. Similarly, a new electrical architecture may be required to power AMDR, which would result in changes to many electrical and machinery control systems and the addition of a fourth large generator. The red team assessment of the Navy's ongoing Flight III technical study found that modifying DDG 51 to accommodate these changes will be challenging with serious design complexity. Since Flight III design work is just in the concept phases, it is currently unknown how the additional cooling and power generating units added to support AMDR will be arranged, or any impact they will have on ship spaces and habitability. For example, the Navy is currently considering five possible cooling unit configurations. Of these, one cannot be arranged within the existing spaces, another will be very difficult to arrange, and three of these options will require significant changes to the arrangements of the chilled water systems. Similarly, all of the options the Navy is considering for possible power generation options will require rearrangement and some impact on other spaces, including encroachment on storage and equipment rooms. Navy officials told us that hybrid electric drive is being researched for Flight III, and the Navy has awarded a number of contracts to study concepts.

Not only can density complicate design of the ship as equipment needs to be rearranged to fit in new items, but Navy data also show that construction of dense vessels tends to be more costly than construction of vessels with more open space. For example, submarine designs are more complicated to arrange and the vessels are more complicated and costly to build than many surface ships. DDG 1000 was designed in part to have reduced density, which could help lower construction costs. According to a 2005 independent study of U.S. naval shipbuilding, any incremental increase in the complexity of an already complex vessel results in a disproportionate increase in work for the shipbuilder, and concluded that cost, technical and schedule risk, and the probability of cost and schedule overrun all increase with vessel density and complexity. The Navy told us that this technology has the ability to generate an additional 1 megawatt of electricity, and thus could potentially obviate the need for an additional generator to support AMDR. However, adding hybrid electric drive would require additional design changes to accommodate the new motors and supporting equipment. Therefore, further adding to the density of DDG 51 to incorporate AMDR is likely to result in higher construction costs and longer construction schedules than on Flight IIA ships....

Costs of the lead Flight III ship will likely exceed current budget estimates. Although the Navy has not yet determined the final configuration for the Flight III ships, regardless of the variant it selects, it will likely need additional funding to procure the lead ship above the level in its current shipbuilding budget. The Navy has estimated $2.6 billion in its fiscal year 2012 budget submission for the lead Flight III ship. However, this estimate may not reflect the significant design and construction challenges that the Navy will face in constructing the Flight III DDG 51s—and the lead ship in particular. In fact, the Navy's most current estimates for a range of notional Flight III options are between $400 million and $1 billion more than current budget estimates, depending on the configuration and equipment of the variant selected....

Further, across the entire flight of 22 ships, the Navy currently estimates Flight III research and development and procurement costs to range from $58.5 billion to $64.1 billion in constant 2012 dollars. However, the Navy estimated in its 2011 long-range shipbuilding plan to Congress that these same 22 ships would cost approximately $50.5 billion in constant 2012 dollars.... depending on the extent of changes to hullform, the Navy may need at least $4.2 billion to $11.4 billion more to procure DDG 51 Flight III ships....

Based on past experience, the Navy's estimates for future DDG 51s will likely increase further as it gains greater certainty over the composition of Flight III and beyond. At the beginning of a program, uncertainty about cost estimates is high. Our work has shown that over time, cost estimates become more certain as the program progresses—and generally increase as costs are better understood and program risks are realized. Recent Navy shipbuilding programs, such as the Littoral Combat Ship program, initially estimated each ship to cost less than $220 million. This estimate has more than doubled as major elements of the ships' design and construction became better understood. In the case of Flight III, the Navy now estimates 3 to 4 additional crew members will be required per Flight III ship to support the IAMD mission and AMDR than it estimated in the earlier Radar/Hull Study. Increases in the cost of Flight III would add further pressure to the Navy's long-range shipbuilding plan. Beginning in 2019, the Navy will face significant constraints on its shipbuilding account as it starts procuring new ballistic missile submarines to replace the current Ohio class. The Navy currently estimates that this program will cost approximately $80.6 billion in procurement alone, with production spanning over a decade....

We recommend that the Secretary of Defense direct the Secretary of the Navy to take the following three actions:...

3. In consultation with MDA and DOD and Navy weapons testers, define an operational testing approach for the Aegis ACB-12 upgrades that includes sufficient simultaneous live-fire testing needed to fully validate IAMD capabilities....

DOD also agreed with our third recommendation on live-fire testing of Aegis ACB-12 upgrades, stating that the Navy and the MDA—working under Office of the Secretary of Defense oversight—are committed to conducting adequate operational testing of ACB-12, but did not offer concrete steps they would take to address our concerns. Moving forward, DOD should demonstrate its commitment to fully validating IAMD capabilities by including robust simultaneous operational live-fire testing of multiple cruise and ballistic missile targets in its Aegis Test and Evaluation Master Plan that is signed by Director, Operational Test and Evaluation.[35]

Flight III DDG-51: Growth Margin

Another issue for Congress is whether the Flight III DDG-51 design would have sufficient growth margin for a projected 35- or 40-year service life. A ship's growth margin refers to its capacity for being fitted over time with either additional equipment or newer equipment that is larger, heavier, or more power-intensive than the older equipment it is replacing, so as to preserve the ship's mission effectiveness. Elements of a ship's growth margin include interior space, weight-carrying capacity, electrical power, cooling capacity (to cool equipment), and ability to accept increases in the ship's vertical center of gravity. Navy ship classes are typically designed so that the first ships in the class will be built with a certain amount of growth margin. Over time, some or all of the growth margin in a ship class may be used up by backfitting additional or newer systems onto existing ships in the class, or by building later ships in the class to a modified design that includes additional or newer systems.

Modifying the DDG-51 design over time has used up some of the design's growth margin. The Flight III DDG-51 would have less of a growth margin than what the Navy would aim to include in a new destroyer design of about the same size.

Supporters of the Navy's proposal to procure Flight III DDG-51s could argue that the ship's growth margin would be adequate because the increase in capability achieved with the Flight III configuration reduces the likelihood that the ship will need much subsequent modification to retain its mission effectiveness over its projected service life. They could also argue that, given technology advances, new systems added to the ship years from now might require no more (and possibly less) space, weight, electrical power, or cooling capacity than the older systems they replace.

Skeptics could argue that there are uncertainties involved in projecting what types of capabilities ships might need to have to remain mission effective over a 35- or 40-year life, and that building expensive new warships with relatively modest growth margins consequently would be imprudent. The Flight III DDG-51's growth margin, they could argue, could make it more likely that the ships would need to be removed from service well before the end of their projected service lives due to an inability to accept modifications needed to preserve their mission effectiveness. Skeptics could argue that it might not be possible to fit the Flight III DDG-51 in the future with an electromagnetic rail gun (EMRG) or a high-power (200 kW to 300 kW) solid state laser (SSL), because the ship would lack the electrical power or cooling capacity required for such a weapon. Skeptics could argue that EMRGs and/or high-power SSLs could be critical to the Navy's ability years from now to affordably counter large numbers of enemy anti-ship cruise missiles (ASCMs) and anti-ship ballistic missiles (ASBMs) that might be fielded by a wealthy and determined adversary. Skeptics could argue that procuring Flight III DDG-51s could delay the point at which EMRGs or high-power SSLs could be introduced into the cruiser-destroyer force, and reduce for many years the portion of the cruiser-destroyer force that could ultimately be backfitted with these weapons. This, skeptics could argue, might result in an approach to AAW

---

[35] Government Accountability Office, *Arleigh Burke Destroyers[:] Additional Analysis and Oversight Required to Support the Navy's Future Surface Combatant Plans*, GAO-12-113, January 2012, pp. 31-38, 45-48, 53.

and BMD on cruisers and destroyers that might ultimately be unaffordable for the Navy to sustain in a competition against a wealthy and determined adversary.[36]

The January 2012 GAO report on DDG-51 acquisition stated:

> The addition of equipment to Flight III adds weight to the ship, and adding the large, heavy AMDR arrays to the deckhouse will also change the ship's center of gravity—defined as the height of the ship's vertical center of gravity as measured from the bottom of the keel, including keel thickness. Weight and center of gravity are closely monitored in ship design due to the impact they can have on ship safety and performance. The Navy has required service life allowances (SLA) for weight and center of gravity for ships to allow for future changes to the ships, such as adding equipment and reasonable growth during the ship's service life—based on historical data—without unacceptable compromises to hull strength, reserve buoyancy, and stability (e.g., tolerance against capsizing). Adding new systems or equipment may require mitigating action such as removing weight (e.g., equipment, combat systems) from the ship to provide enough available weight allowance to add desired new systems or equipment. A reduced center of gravity may require mitigation such as adding additional weight in the bottom of the ship to act as ballast, though this could also reduce the available weight allowance. These changes all require redesign which can increase costs, and this design work and related costs can potentially recur over the life of the ship.
>
> The Navy is considering a range of design options to deal with adding AMDR and its supporting power and cooling equipment. None of the DDG 51 variants under consideration as part of an ongoing Navy study meet Navy SLA requirements of 10 percent of weight and 1 foot of center of gravity for surface combatants.... several variants provide less than half of the required amounts.
>
> The Navy has determined that only by completely changing the material of the entire fore and aft deckhouses and the helicopter hangars to aluminum or composite as well as expanding the overall dimensions of the hull (especially the width, or beam) can the full SLA be recovered for a Flight III with a 14-foot AMDR. Though a decision has not yet been made, at this time Navy officials do not believe that a composite or aluminum deckhouse will be used. The Navy also told us that removing combat capability from DDG 51 may be required in an effort to manage weight after adding AMDR, effectively reducing the multimission functionality of the class. Navy officials stated that SLA has not always been required, and that this allowance is included in designs to eventually be consumed. They pointed to other classes of ships that were designed with less than the required SLA margins and that have performed adequately. However, as shown in Table 10, our analysis of the data indicates that these ships have faced SLA-related issues.
>
> According to Navy data, delivery weight of DDG 51s has gotten considerably heavier over the course of building the class, with current 51s weighing approximately 700-900 long tons (a measure of ship displacement) more than the first DDG 51s. Further, while the current DDG 51s all can accept both an increase in weight or rise in the center of gravity, the ships are already below the required center of gravity allowances, though Navy officials told us that this could be corrected with ballasting if the Navy opted to fund the change. In commenting on the ongoing Navy study, the independent red team identified reduced SLA as a significant concern for Flight III, and noted that if the Navy does not create a larger hull form for Flight III, any future ship changes will be significantly constrained....

---

[36] For more on potential shipboard lasers, see CRS Report R41526, *Navy Shipboard Lasers for Surface, Air, and Missile Defense: Background and Issues for Congress*, by Ronald O'Rourke.

Officials told us that a major consideration in the future will be electrical power. While Flight III will most likely not leverage technologies developed as part of the DDG 1000 program because of DDG 51's design constraints, Navy officials stated that [the projected future] Flight IV [version of the ship] may carry some form of the integrated power system developed for DDG 1000. The Navy examined the use of the integrated power system for Flight III in the Flight Upgrade Study, but found that it was not currently viable due to current component technology. The constrained nature of Flight III will likely limit the ability of the Navy to add future weapon technologies to these ships—such as an electromagnetic rail gun or directed energy weapons as these technologies mature—unless the Navy wants to remove current weapon systems. For example, the ongoing Navy Flight Upgrade Study examined an option to add a small rail gun by removing the ship's main 5-inch gun and the forward 32-cell missile launcher system. It is unknown when these future technologies may be used.[37]

Flight IIA DDG-51: Schedule Risk

Another issue for Congress concerns schedule risk for recently procured Flight IIA DDG-51s. The January 2012 GAO report on DDG-51 acquisition stated:

> While the shipbuilders' planned production schedules are generally in line with past shipyard performance, the delivery schedule for the first [Flight IIA] restart ship (DDG 113) may be challenging because of a significant upgrade in the Aegis combat system, where major software development efforts are under way and a critical component has faced delays. Although the Navy plans to install and test this upgrade on an older DDG 51 (DDG 53) prior to installation on DDG 113, delays in these efforts could pose risks to a timely delivery in support of DDG 113 and ability to mitigate risk. If this occurs, the Navy may need additional time to identify, analyze, and work to resolve problems with the combat system—adding pressure to the schedule for DDG 113....

> The schedules [for building the Flight IIA ships], while in line with past performance, are contingent on achieving an optimum build sequence, meaning the most efficient schedule for constructing a ship, including building the ship from the bottom up and installing ship systems before bulkheads have been built and when spaces are still easily accessible. Shipbuilders generate specific dates for when systems need to arrive at the shipyard in order to take advantage of these efficiencies. According to shipyard officials, approximately 10 percent to 12 percent of the suppliers for the restart ships will be new vendors. Some key pieces of equipment—like the main reduction gear, the machinery control system, and the engine controllers—will now be government-furnished equipment, meaning that the Navy will be responsible for ensuring an on-time delivery to the shipyard, not the shipbuilder. For the main reduction gear, the Navy is now contracting with a company that bought the gear production line from the past supplier, and while this supplier builds reduction gears for San Antonio class ships, it does not have experience building DDG 51 main reduction gears. An on-time delivery of this key component is particularly important to the schedule because it is installed early in the lower sections of the ship. A delay in a main reduction gear could result in a suboptimal build sequence as the shipbuilder has to restructure work to leave that space open until the gear arrives. The Defense Contract Management Agency reports production of the first gear ship set is progressing well, and that Navy officials are tracking the schedule closely.

---

[37] Government Accountability Office, *Arleigh Burke Destroyers[:] Additional Analysis and Oversight Required to Support the Navy's Future Surface Combatant Plans*, GAO-12-113, January 2012, pp. 38-41, 45, 52.

A major change for the restart ships is a significant upgrade to the Aegis combat system currently under way. This upgrade, known as Advanced Capability Build 12 (ACB 12), will be retrofitted on some of the current fleet of DDG 51s (starting with DDG 53); following DDG 53, the upgrade will also be installed on the restart ships (starting with DDG 113). The retrofit on DDG 53 will provide the Navy with a risk mitigation opportunity, since any challenges or problems can be identified and resolved prior to installation on DDG 113. The Navy believes this is the most complex Aegis upgrade ever undertaken and will enable the combat system to perform limited IAMD for the first time. This upgrade will also move the Navy towards a more open architecture combat system, meaning that there will be a reduction of proprietary software code and hardware so that more elements can be competitively acquired in the future. To date, Lockheed Martin maintains intellectual property rights over some Aegis components. ACB 12 requires both software and hardware changes, and consists of three related development efforts: (1) development of a multimission signal processor (MMSP), (2) changes to the ballistic missile suite (BMD 5.0), and (3) changes to the Aegis combat system core. While the Navy manages the development of MMSP and ACB 12, MDA manages the development of BMD 5.0. Table 8 describes each of the three efforts.

While the Navy has made significant progress in developing the components of ACB 12, MMSP is proving more difficult than estimated and is currently 4 months behind schedule, with $10 million in cost growth realized and an additional $5 million projected. A substantial amount of software integration and testing remains before MMSP can demonstrate full capability and is ready for installation on DDG 53—and later DDG 113. While all of the software has been developed, only 28 percent of the eight software increments have been integrated and tested. The integration phase is typically the most challenging in software development, often requiring more time and specialized facilities and equipment to test software and fix defects. According to the Navy, the contractor underestimated the time and effort required to develop and integrate the MMSP software. In December 2010, MMSP was unable to demonstrate planned functionality for a radar test event due to integration difficulties, and MMSP more recently experienced software problems during radar integration which resulted in schedule delays. In response, the contractor implemented a recovery plan, which included scheduling additional tests and replanning the remaining work to improve system stability. However, the recovery plan compresses the time allocated for integrating MMSP with the rest of the combat system from 10 months to 6 months.

In order to meet schedule goals and mitigate software development risk in the near term, the contractor also moved some development of MMSP capability to future builds. However, this adds pressure to future development efforts and increases the probability of defects and integration challenges being realized late in the program. The contractor already anticipates a 126 percent increase in the number of software defects that it will have to correct over the next year, indicating the significant level of effort and resources required for the remaining development. According to the program office, the high level of defects Each defect takes time to identify and correct, so a high level of defects could result in significant additional work and potentially further delays if the contractor cannot resolve the defects as planned. The Navy believes the schedule risk associated with this increase is understood and anticipates no further schedule impacts. However, the Defense Contract Management Agency, which is monitoring the combat system development for the Navy, has characterized the MMSP schedule as high risk.

.... the Navy will not test ACB 12's IAMD capabilities with combined live ballistic and cruise missile tests until after it certifies the combat system. Certification is an assessment of the readiness and safety of ACB 12 for operational use including the ability to perform Aegis ship missions. The Navy and MDA plan to determine future opportunities for additional testing to prove the system. The Navy plans to leverage a first quarter fiscal year 2015 test that MDA does not actually characterize as an IAMD test to demonstrate IAMD capabilities.

The Navy initially planned to test the combat system's IAMD tracking capability during a BMD test event to occur by third quarter fiscal year 2013. The test—tracking and simulated engagement of BMD and air warfare targets—would have provided confidence prior to certification of ACB 12 that the software worked as intended. However, this event was removed from the test schedule The Navy now plans to test tracking and simulated IAMD engagement capability during a BMD test event in third quarter 2014. According to the Navy, this is the earliest opportunity for sea-based testing of the ACB 12 upgrade installed on DDG 53. This event will help demonstrate functionality and confidence in the system, but only allows five months between the test and certification of the system to resolve any problems that may be identified during testing. The Navy and MDA plan on conducting a live ballistic missile exercise in second quarter fiscal year 2014, this will only test the combat system's BMD capability, not IAMD. Consequently, the Navy will certify that the combat system is mission ready without validating with live ballistic and cruise missile targets that it can perform the IAMD mission. The first IAMD test with live targets is not scheduled until first quarter fiscal year 2015.

Delays in MMSP could also lead to concurrence between final software integration and the start of ACB 12 installation on DDG 53. Although the Navy has stated that the contractor is currently on schedule, if the contractor is unable to resolve defects according to plan, Aegis Light-Off (when the combat system is fully powered on for the first time) on DDG 53 could slip or the test period could move closer to the start of installation on DDG 113, which could limit risk mitigation opportunities. Contractor officials told us that they plan to deliver the combat system hardware to the shipyard for installation on DDG 113 in May 2013. While the Navy believes the current schedule allows time for the Navy and contractor to remedy any defects or problems found with ACB 12 before it is scheduled to be installed on DDG 113, we have previously reported that concurrent development contributes to schedule slips and strains resources required to develop, integrate, test, and rework defects, which could encroach into this buffer.

Additionally, if DDG 53 is not available when currently planned to begin its upgrade, this process could also be delayed. DDG 53's upgrade schedule already slipped from May 2012 to September 2012, and any significant shifts could mean further schedule compression, or if it slipped past the start of installation on DDG 113 this new-construction ship could become the ACB 12 test bed, which would increase risk.[38]

Options for Congress

In general, options for Congress concerning destroyer acquisition include the following:

- approving, rejecting, or modifying the Navy's procurement, advance procurement, and research and development funding requests for destroyers and their associated systems (such as the AMDR);

- establishing conditions for the obligation and expenditure of funding for destroyers and their associated systems; and

---

[38] Government Accountability Office, *Arleigh Burke Destroyers[:] Additional Analysis and Oversight Required to Support the Navy's Future Surface Combatant Plans*, GAO-12-113, January 2012, pp. 20, 24-29. See also Megan Eckstein, "DDG Restart Ready For Multiyear After Supply Chain Problems Addressed," *Inside the Navy*, January 16, 2012.

- holding hearings, directing reports, and otherwise requesting information from DOD on destroyers and their associated systems.

In addition to these general options, below are some additional acquisition options relating to destroyers that Congress may wish to consider.

Adjunct Radar Ship

The Navy canceled the CG(X) cruiser program in favor of developing and procuring Flight III DDG-51s reportedly in part on the grounds that the Flight III destroyer would use data from off-board sensors to augment data collected by its AMDR.[39] If those off-board sensors turn out to be less capable than the Navy assumed when it decided to cancel the CG(X) in favor of the Flight III DDG-51, the Navy may need to seek other means for augmenting the data collected by the Flight III DDG-51's AMDR.

One option for doing this would be to procure an adjunct radar ship—a non-combat ship equipped with a large radar that would be considerably more powerful than the Flight III DDG-51's AMDR. The presence in the fleet of a ship equipped with such a radar could significantly improve the fleet's AAW and BMD capabilities. The ship might be broadly similar to (but perhaps less complex and less expensive than) the new Cobra Judy Replacement missile range instrumentation ship (**Figure 2**),[40] which is equipped with two large and powerful radars, and

---

[39] Amy Butler, "STSS Prompts Shift in CG(X) Plans," *Aerospace Daily & Defense Report*, December 11, 2009: 1-2.

[40] As described by DOD,

> The COBRA JUDY REPLACEMENT (CJR) program replaces the capability of the current United States Naval Ship (USNS) Observation Island (OBIS), its COBRA JUDY radar suite, and other mission essential systems. CJR will fulfill the same mission as the current COBRA JUDY/OBIS. CJR will collect foreign ballistic missile data in support of international treaty verification.
>
> CJR represents an integrated mission solution: ship, radar suite, and other Mission Equipment (ME). CJR will consist of a radar suite including active S-Band and X-Band Phased Array Radars (PARs), weather equipment, and a Mission Communications Suite (MCS). The radar suite will be capable of autonomous volume search and acquisition. The S-Band PAR will serve as the primary search and acquisition sensor and will be capable of tracking and collecting data on a large number of objects in a multi-target complex. The X-Band PAR will provide very high-resolution data on particular objects of interest....
>
> The OBIS replacement platform, USNS Howard O. Lorenzen (Missile Range Instrumentation Ship (T-AGM) 25), is a commercially designed and constructed ship, classed to American Bureau of Shipping standards, certified by the U.S. Coast Guard in accordance with Safety of Life at Sea, and in compliance with other commercial regulatory body rules and regulations, and other Military Sealift Command (MSC) standards. The ship will be U.S. flagged, operated by a Merchant Marine or MSC Civilian Mariner crew, with a minimum of military specifications. The ship is projected to have a 30-year operating system life-cycle.
>
> The U.S. Navy will procure one CJR for the U.S. Air Force using only Research, Development, Test and Evaluation funding. CJR will be turned over to the U.S. Air Force at Initial Operational Capability for all operations and maintenance support....
>
> Program activities are currently focused on installation and final integration of the X and S-band radars onto the ship at Kiewit Offshore Services (KOS) following completion of radar production and initial Integration and Test (I&T) at Raytheon and Northrop Grumman (NG). Raytheon and its subcontractors have completed I&T of the X-band radar and X/S ancillary equipment at KOS. The S-band radar arrived at KOS on February 19, 2011. The United States Naval Ship (USNS) Howard O. Lorenzen (Missile Range Instrumentation Ship (T-AGM) 25) completed at-sea Builder's Trials (BT) in March 2011. The ship is expected to depart VT Halter Marine (VTHM) and arrive at KOS in the third quarter of Fiscal Year 2011 (3QFY11).

(continued...)

---

which has an estimated total acquisition cost of about $1.7 billion.[41] One to a few such adjunct radar ships might be procured, depending on the number of theaters to be covered, requirements for maintaining forward deployments of such ships, and their homeporting arrangements. The ships would have little or no self-defense capability and would need to be protected in threat situations by other Navy ships.

Figure 2. Cobra Judy Replacement Ship

Source: Naval Research Laboratory (http://www.nrl.navy.mil/PressReleases/2010/image1_74-10r_hires.jpg, accessed on April 19, 2011).

Flight III DDG-51 With Increased Capabilities

Another option would be to design the Flight III DDG-51 to have greater capabilities than what the Navy is currently envisioning. Doing this might well require the DDG-51 hull to be lengthened—something that the Navy currently does not appear to be envisioning for the Flight III design. Navy and industry studies on the DDG-51 hull design that were performed years ago suggested that the hull has the potential for being lengthened by as much as 55 feet to accommodate additional systems. Building the Flight III DDG-51 to a lengthened configuration could make room for additional power-generation and cooling equipment, additional vertical launch system (VLS) missile tubes, and larger growth margins. It might also permit a redesign of the deckhouse to support a larger and more capable version of the AMDR than the 12- or 14-foot diameter version currently planned for the Flight III DDG-51. Building the Flight III DDG-51 to a lengthened configuration would increase its develop cost and its unit procurement cost. The increase in unit procurement cost could reduce the number of Flight III DDG-51s that the Navy could afford to procure without reducing funding for other programs.

(...continued)

(Department of Defense, *Selected Acquisition Report (SAR), Cobra Judy Replacement*, December 31, 2010, pp. 3-5.)

[41] Department of Defense, *Selected Acquisition Report (SAR), Cobra Judy Replacement*, December 31, 2010, p. 13.

DDG-1000 Variant With AMDR

Another option would be to design and procure a version of the DDG-1000 destroyer that is equipped with the AMDR and capable of BMD operations. Such a ship might be more capable in some regards than the Flight III DDG-51, but it might also be more expensive to develop and procure. An AMDR-equipped, BMD-capable version of the DDG-1000 could be pursued as either a replacement for the Flight III DDG-51 or a successor to the Flight III DDG-51 (after some number of Flight III DDG-51s were procured). A new estimate of the cost to develop and procure an AMDR-equipped, BMD-capable version of the DDG-1000 might differ from the estimate in the 2009 destroyer hull/radar study due to the availability of updated cost information for building the current DDG-1000 design.

New-Design Destroyer

Another option would be to design and procure a new-design destroyer that is intermediate in size between the DDG-51 and DDG-1000 designs, equipped with the AMDR, and capable of BMD operations. This option could be pursued as either a replacement for the Flight III DDG-51 or a successor to the Flight III DDG-51 (after some number of Flight III DDG-51s were procured). Such a ship might be designed with the following characteristics:

- a version of the AMDR that is larger than the one envisioned for the Flight III DDG-51, but smaller than the one envisioned for the CG(X);

- enough electrical power and cooling capacity to permit the ship to be backfitted in the future with an EMRG or high-power SSL;

- more growth margin than on the Flight III DDG-51;

- producibility features for reducing construction cost per ton that are more extensive than those on the DDG-51 design;

- automation features permitting a crew that is smaller than what can be achieved on a Flight III DDG-51, so as to reduce crew-related life-cycle ownership costs;

- physical open-architecture features that are more extensive than those on the Flight III DDG-51, so as to reduce modernization-related life-cycle ownership costs;

- no technologies not already on, or being developed for, other Navy ships, with the possible exception of technologies that would enable an integrated electric drive system that is more compact than the one used on the DDG-1000; and

- DDG-51-like characteristics in other areas, such as survivability, maximum speed, cruising range, and weapons payload.

Such a ship might have a full load displacement of roughly 11,000 to 12,000 tons, compared to about 10,000 tons for the Flight III DDG-51, 15,000 or more tons for an AAW/BMD version of the DDG-1000, and perhaps 15,000 to 23,000 tons for a CG(X).[42]

---

[42] The cost and technical risk of developing the new destroyer's hull design could be minimized by leveraging, where possible, existing surface combatant hull designs. The cost and technical risk of developing its combat system could be minimized by using a modified version of the DDG-51 or DDG-1000 combat system. Other development costs and (continued...)

Legislative Activity for FY2013

FY2013 Funding Request

The Navy's proposed FY2013 budget requests $3,048.6 million to complete the procurement funding for the two DDG-51s scheduled for procurement in FY2013. The Navy estimates the total procurement cost of these ships at $3,149.4 million, and the ships have received $100.7 million in prior-year advance procurement (AP) funding. The FY2013 budget also requests $466.3 million in AP funding for DDG-51s to be procured in future fiscal years. Much of this AP funding is for Economic Order Quantity (EOQ) procurement of selected components of the nine DDG-51s to be procured under the proposed FY2013-FY2017 MYP arrangement. The Navy's proposed FY2013 budget also requests $669.2 million in procurement funding to help complete procurement costs for three Zumwalt (DDG-1000) class destroyers procured in FY2007-FY2009, and $223.6 million in research and development funding for the AMDR. The funding request for the AMDR is contained in the Navy's research and development account in Project 3186 ("Air and Missile Defense Radar") of Program Element (PE) 0604501N ("Advanced Above Water Sensors").

FY2013 National Defense Authorization Act (H.R. 4310/S. 3254)

House

Section 125 of H.R. 4310 as reported by the House Armed Services Committee (H.Rept. 112-479 of May 11, 2012) states:

> SEC. 125. MULTIYEAR PROCUREMENT AUTHORITY FOR ARLEIGH BURKE-CLASS DESTROYERS AND ASSOCIATED SYSTEMS.
>
> (a) Authority for Multiyear Procurement- In accordance with section 2306b of title 10, United States Code, the Secretary of the Navy may enter into a multiyear contract, beginning with the fiscal year 2013 program year, for the procurement of not more than 10 Arleigh Burke-class guided missile destroyers, including the Aegis weapon systems, MK 41 vertical launching systems, and commercial broadband satellite systems associated with such vessels.

(...continued)

risks for the new destroyer would be minimized by using no technologies not already on, or being developed for, other Navy ships (with the possible exception of some integrated electric drive technologies). Even with such steps, however, the cost and technical risk of developing the new destroyer would be greater than those of the Flight III DDG-51. The development cost of the new destroyer would likely be equivalent to the procurement cost of at least one destroyer, and possibly two destroyers.

The procurement cost of the new destroyer would be minimized by incorporating producibility features for reducing construction cost per ton that are more extensive than those on the Flight III DDG-51. Even with such features, the new destroyer would be more expensive to procure than the Flight III DDG-51, in part because the Flight III DDG-51 would leverage many years of prior production of DDG-51s. In addition, the new destroyer, as a new ship design, would pose more risk of procurement cost growth than would the Flight III DDG-51. The procurement cost of the new destroyer would nevertheless be much less than that of the CG(X), and might, after the production of the first few units, be fairly close to that of the Flight III DDG-51.

(b) Authority for Advance Procurement- The Secretary of the Navy may enter into a contract, beginning in fiscal year 2013, for advance procurement associated with the vessels and systems for which authorization to enter into a multiyear procurement contract is provided under subsection (a).

(c) Condition for Out-year Contract Payments- A contract entered into under subsection (a) shall provide that any obligation of the United States to make a payment under the contract for a fiscal year after fiscal year 2013 is subject to the availability of appropriations or funds for that purpose for such later fiscal year.

Regarding Section 125, H.Rept. 112-479 states: "For many years, this class of ships was efficiently procured through multiyear procurement contracts, until the restart of production. The DDG–51 Flight IIA possesses a stable design and the committee supports the budget request to continue DDG–51 production through the Future Years Defense Program." (Page 50)

The report recommends approving the Navy's request for FY2013 procurement funding for the DDG-51 program, increasing by $115 million the Navy's request for FY2013 advance procurement funding for the DDG-51 program, and approving the Navy's request for FY2013 procurement funding for the DDG-1000 program. (Page 375) The report recommends approving the Navy's request for FY2013 research and development funding for PE0604501N (Advanced Above Water Sensors). (Page 421)

The report states:

> In the fiscal year 2013 budget request, the Department of the Navy has requested authority to begin a multi-year program for nine DDG–51 Arleigh Burke-class destroyers. Elsewhere in this Act, the committee includes a provision [Section 125] that would authorize the Secretary of the Navy to award a contract for a multiyear procurement of up to 10 destroyers. In fiscal year 2016, the Navy intends to start procuring Block III DDG–51 destroyers. This block will incorporate the advanced Air and Missile Defense Radar (AMDR), which is currently being competitively evaluated. The committee views AMDR as essential to pacing the air and missile threat. The Navy has stated that the DDG–51 hull is sufficient to accommodate the increased power generation and cooling requirements that AMDR will need, yet the committee still views this as an area of risk. (Page 35)

The report also states:

> *Shipbuilding Material Comparison*
>
> In a recent article published in "Inside the Navy", it was reported that, "superstructure cracking in several classes of surface combatants is being addressed, but in some cases is proving costly". The committee is aware that three materials have been used in the deckhouses of surface combatants: steel, aluminum, and most recently for the deckhouse of the DDG–1000 Zumwalt class, composite material.
>
> The committee is also aware that there is a cost differential in both up-front procurement and production and in lifecycle maintenance cost for these materials. The next opportunity that the Navy will have to influence a design will be with Flight III of the DDG–51 Arleigh-Burke destroyers. The committee directs the Secretary of the Navy to provide a report to the congressional defense committees with delivery of the fiscal year 2014 budget request, comparing the estimated construction costs for a deckhouse made of each of the three materials, or even a possible hybrid of two or all three, and then compares the estimated lifecycle costs for the designed life of the ship. (Pages 66-67)

Senate

Section 125 of S. 3254 as reported by the Senate Armed Services Committee (S.Rept. 112-173 of June 4, 2012) states:

> SEC. 125. MULTIYEAR PROCUREMENT AUTHORITY FOR ARLEIGH BURKE CLASS DESTROYERS AND ASSOCIATED SYSTEMS.
>
> (a) Authority for Multiyear Procurement- Subject to section 2306b of title 10, United States Code, the Secretary of the Navy may enter into multiyear contracts, beginning with the fiscal year 2013 program year, for the procurement of up to 10 Arleigh Burke class Flight IIA guided missile destroyers, as well as the AEGIS Weapon Systems, MK 41 Vertical Launching Systems, and Commercial Broadband Satellite Systems associated with those vessels.
>
> (b) Authority for Advance Procurement- The Secretary may enter into one or more contracts, beginning in fiscal year 2013, for advance procurement associated with the vessels and systems for which authorization to enter into a multiyear procurement contract is provided under subsection (a).
>
> (c) Condition for Out-year Contract Payments- A contract entered into under subsection (a) shall provide that any obligation of the United States to make a payment under the contract for a fiscal year after fiscal year 2013 is subject to the availability of appropriations or funds for that purpose for such later fiscal year.

Regarding Section 125, S.Rept. 112-173 states:

> **Multiyear procurement authority for Arleigh Burke-class destroyers and associated systems (sec. 125)**
>
> The committee recommends a provision [Section 125] that would authorize the Secretary of the Navy to buy up to 10 Arleigh Burke-class Flight IIA destroyers under a multiyear procurement contract. This would be the third multiyear contract for the Arleigh Burke-class program. The Navy estimates that the previous two multiyear procurement contracts (fiscal years 1998–2001 and fiscal years 2002–2005) achieved savings of greater than $1.0 billion, as compared to annual procurements. For the third contract (for fiscal years 2013–2017), the Navy is estimating that the expected savings will be 8.7 percent, or in excess of $1.5 billion, for the multiyear approach as compared to annual procurement contracts.
>
> While the Navy's shipbuilding plan currently provides for only nine Arleigh Burke-class destroyers during the period of the planned multiyear contract, the committee understands from the Navy that competition between the two shipyards in fiscal year 2011 and 2012 has led to significant savings in the program compared to the original budget request. The Navy program office believes that competition for the multiyear contract starting in fiscal year 2013 could also yield additional savings, and that the sum total of those savings might be sufficient to purchase an additional destroyer in fiscal year 2014. The committee is recommending approval of a multiyear authority for up to 10 ships with the prospect that the Navy may be able to combine the savings from fiscal years 2011, 2012, and 2013 and buy an additional destroyer, which is consistent with congressional authorization in section 2308 of title 10, United States Code, for the Secretary to buy-to-budget.
>
> The committee believes that continued production of Arleigh Burke-class destroyers is critical to provide required forces for seabased ballistic missile defense (BMD) capabilities. The Navy envisions that, if research and development activities yield an improved radar suite

and combat systems capability, they would like to install those systems on the destroyers in fiscal years 2016 and 2017, at which time the designation for those destroyers would be Flight III. Should the Navy decide to move forward with the integration of an engineering change proposal (ECP) to incorporate a new BMD capable radar and associated support systems during execution of this multiyear procurement, the Secretary of the Navy shall submit a report to the congressional defense committees, no later than with the budget request for the year of contract award of such an ECP. The report will contain a description of the final scope of this ECP, as well as the level of maturity of the new technology to be incorporated on the ships of implementation and rationale as to why the maturity of the technology and the capability provided justify execution of the change in requirements under that ECP during the execution of a multiyear procurement contract. (Pages 13-14)

The report recommends approving the Navy's request for FY2013 procurement funding for the DDG-51 program, the Navy's request for FY2013 advance procurement funding for the DDG-51 program, and the Navy's request for FY2013 procurement funding for the DDG-1000 program. (Page 325) The report recommends approving the Navy's request for FY2013 research and development funding for PE0604501N (Advanced Above Water Sensors). (Page 375)

FY2013 DOD Appropriations Bill (H.R. 5856)

House

Section 8010 of H.R. 5856 as reported by the House Appropriations Committee (H.Rept. 112-493 of May 25, 2012) provides authority for MYP arrangements for certain procurement programs, including the DDG-51 program.

H.Rept. 112-493 recommends increasing by a net total of $987.97 million the Navy's request for FY2013 procurement funding for the DDG-51 program, with the net increase consisting of an increase of $1 billion to facilitate the procurement of an additional DDG-51 in FY2013, and reductions of $10.214 million for "EXCOMM equipment cost growth" and $1.816 million for "CIWS [close-in weapon system] hardware cost growth." The report recommends approving the Navy's requests for FY2013 advance procurement funding for the DDG-51 program and for FY2013 procurement funding for the DDG-1000 program. (Pages 156 and 157) The report recommends approving the Navy's request for FY2013 research and development funding for PE0604501N (Advanced Above Water Sensors). (Page 223)

The report states:

> The decision to defer the procurement of a [Virginia-class] submarine [from FY2014 to FY2018] and a [DDG-51] destroyer [from FY2014 to FY2016] is both confusing and concerning, especially the submarine....
>
> The Committee believes the Navy recognizes the need to fund another destroyer and submarine in fiscal year 2014 since the Navy has approached the Committee with various plans and schemes to attempt to restore these ships to fiscal year 2014....
>
> The Committee understands the constraints of the fiscal year 2014 budget, but to give up two highly prized combatants, and fund instead a vessel for a mission that can be (and has been) satisfied with existing ships, then attempt to restore those combatants through funding gimmicks in violation of the Department's own financial regulations is deeply troubling. The Committee firmly believes that a strong Navy shipbuilding program is absolutely essential

for the Nation's security but will not mortgage the Nation's future to accomplish it. Accordingly, the recommendation provides an additional $1,000,000,000 above the request for the procurement of an additional DDG–51 guided missile destroyer. The Secretary of the Navy is directed to use this funding as part of the DDG–51 multiyear procurement planned for fiscal years 2013 through 2017 in order to achieve a lower cost and provide a more stable production base for the duration of the DDG–51 multiyear procurement. (Pages 158-159)

Senate

Section 8010 of H.R. 5856 as reported by the House Appropriations Committee (S.Rept. 112-196 of August 2, 2012) provides authority for MYP arrangements for certain procurement programs, including "up to 10 DDG-51 Arleigh Burke class Flight IIA guided missile destroyers, as well as the AEGIS Weapon Systems, MK 41 Vertical Launching Systems, and Commercial Broadband Satellite Systems associated with those vessels."

S.Rept. 112-196 recommends increasing by $1 billion the Navy's request for FY2013 procurement funding for the DDG-51 program, with the increase being for the procurement of an additional DDG-51 in FY2013. The report recommends approving the Navy's requests for FY2013 advance procurement funding for the DDG-51 program and for FY2013 procurement funding for the DDG-1000 program. (Pages 124 and 125) The report recommends approving the Navy's FY2013 funding request for research and development work on the Air and Missile Defense Radar (AMDR), and recommends transferring this funding from the Advanced Above Water Sensors program element to a newly created program element in the Navy's research and development account that would be specifically for the AMDR. (Page 185, lines 113 and 113A, and page 190, lines 113 and 113X) The report states:

> For the Department of the Navy, the Committee does not concur with the recommendation to prematurely retire nine Navy ships and provides over $2,300,000,000 to man, operate, equip, and modernize these ships. In addition, the Committee provides over $770,000,000 for advance procurement of a tenth Virginia-class submarine, $1,000,000,000 for an additional DDG–51 destroyer, and $263,255,000 for advance procurement of an amphibious warship. These funds were not included in the budget request, but the Committee believes these ships are crucial to supporting our Navy's global requirements, particularly as the U.S. military shifts its focus to the Pacific. (Page 7)

The report also states:

> The Committee is concerned with the apparent disconnect between the Navy's publicly stated priorities and the Navy's fiscal year 2013 shipbuilding budget submission which, as compared to the fiscal year 2012 plan, reduces planned ship procurement for the next 5 years by 16 ships and eliminates funding for one Virginia class attack submarine, one amphibious ship, and three oilers. In addition, the Navy's budget request omits funding for an additional Arleigh Burke class guided missile destroyer contrary to the Navy's force structure goals. The Committee is concerned that these proposed adjustments will drive up costs for multiple classes of ships and negatively impact the industrial base, almost certainly leading to additional cost growth in shipbuilding....
>
> In addition, the fiscal year 2013 budget request includes a proposal to procure nine Arleigh Burke class guided missile destroyers at a rate of two ships per year for the next 5 years with the exception of fiscal year 2014, where the Navy has only budgeted for one Arleigh Burke class guided missile destroyer. The Committee has been informed by the Navy that a tenth destroyer could be added to this multiyear procurement at a significantly lower ship unit

cost. According to the Navy, those costs could be partially offset with savings generated from previous competitive procurements, with the balance of funding required in the fiscal year 2014 shipbuilding budget request. Recognizing the fiscal pressures on the Navy's fiscal year 2014 shipbuilding budget, the Committee recommends taking advantage of these significant cost savings now and recommends an additional $1,000,000,000 in fiscal year 2013 to fully fund a tenth Arleigh Burke class guided missile destroyer within the next multiyear procurement.

The Committee notes that within this multiyear procurement, the Navy intends to include next generation Flight III Arleigh Burke class guided missile destroyers with an improved radar suite and combat systems capability as an Engineering Change Proposal [ECP]. Since the radar is still in development, acquisition authorities within the Department have yet to be established, and the ECP has no defined scope and associated cost estimates, the Committee finds it premature to request authority for this ECP within the multiyear procurement at this time. Therefore, the Committee provides multiyear procurement authority only for Flight IIA Arleigh Burke class guided missile destroyers, as authorized in S. 3254, the National Defense Authorization Act for Fiscal Year 2013, as reported. (Pages 125-126)

Appendix A. Additional Background Information on DDG-1000 Program

This appendix presents additional background information on the DDG-1000 program.

Program Origin

The program known today as the DDG-1000 program was announced on November 1, 2001, when the Navy stated that it was replacing a destroyer-development effort called the DD-21 program, which the Navy had initiated in the mid-1990s, with a new Future Surface Combatant Program aimed at developing and acquiring a family of three new classes of surface combatants:[43]

- **a destroyer called DD(X)** for the precision long-range strike and naval gunfire mission;

- **a cruiser called CG(X)** for the air defense and ballistic missile mission; and

- **a smaller combatant called the Littoral Combat Ship (LCS)** to counter submarines, small surface attack craft (also called "swarm boats"), and mines in heavily contested littoral (near-shore) areas.[44]

On April 7, 2006, the Navy announced that it had redesignated the DD(X) program as the DDG-1000 program. The Navy also confirmed in that announcement that the first ship in the class, DDG-1000, is to be named the *Zumwalt*, in honor of Admiral Elmo R. Zumwalt, the Chief of Naval operations from 1970 to 1974. The decision to name the first ship after Zumwalt was made by the Clinton Administration in July 2000, when the program was still called the DD-21 program.[45]

New Technologies

The DDG-1000 incorporates a significant number of new technologies, including a wave-piercing, tumblehome hull design for reduced detectability,[46] a superstructure made partly of large sections of composite (i.e., fiberglass-like) materials rather than steel or aluminum, an integrated

---

[43] The DD-21 program was part of a Navy surface combatant acquisition effort begun in the mid-1990s and called the SC-21 (Surface Combatant for the 21st Century) program. The SC-21 program envisaged a new destroyer called DD-21 and a new cruiser called CG-21. When the Navy announced the Future Surface Combatant Program in 2001, development work on the DD-21 had been underway for several years, while the start of development work on the CG-21 was still years in the future. The current DDG-1000 destroyer CG(X) cruiser programs can be viewed as the descendants, respectively, of the DD-21 and CG-21. The acronym SC-21 is still used in the Navy's research and development account to designate the line item (i.e., program element) that funds development work on both the DDG-1000 and CG(X).

[44] For more on the LCS program, see CRS Report RL33741, *Navy Littoral Combat Ship (LCS) Program: Background and Issues for Congress*, by Ronald O'Rourke.

[45] For more on Navy ship names, see CRS Report RS22478, *Navy Ship Names: Background for Congress*, by Ronald O'Rourke.

[46] A tumblehome hull slopes inward, toward the ship's centerline, as it rises up from the waterline, in contrast to a conventional flared hull, which slopes outward as it rises up from the waterline.

---

electric-drive propulsion system,[47] a total-ship computing system for moving information about the ship, automation technologies enabling its reduced-sized crew, a dual-band radar, a new kind of vertical launch system (VLS) for storing and firing missiles, and two copies of a 155mm gun called the Advanced Gun System (AGS). The AGS is to fire a new rocket-assisted 155mm shell, called the Long Range Land Attack Projectile (LRLAP), to ranges of more than 60 nautical miles. The DDG-1000 can carry 600 LRLAP rounds (300 for each gun), and additional rounds can be brought aboard the ship while the guns are firing, creating what Navy officials call an "infinite magazine."

Planned Quantity

When the DD-21 program was initiated, a total of 32 ships was envisaged. In subsequent years, the planned total for the DD(X)/DDG-1000 program was reduced to 16 to 24, then to 7, and finally to 3.

Construction Shipyards

Under a DDG-1000 acquisition strategy approved by the Under Secretary of Defense for Acquisition, Technology, and Logistics (USD AT&L) on February 24, 2004, the first DDG-1000 was to have been built by HII/Ingalls, the second ship was to have been built by GD/BIW, and contracts for building the first six were to have been equally divided between HII/Ingalls[48] and GD/BIW.

In February 2005, Navy officials announced that they would seek approval from USD AT&L to instead hold a one-time, winner-take-all competition between HII/Ingalls and GD/BIW to build all DDG-1000s. On April 20, 2005, the USD AT&L issued a decision memorandum deferring this proposal, stating in part, "at this time, I consider it premature to change the shipbuilder portion of the acquisition strategy which I approved on February 24, 2004."

Several Members of Congress also expressed opposition to Navy's proposal for a winner-take-all competition. Congress included a provision (§1019) in the Emergency Supplemental Appropriations Act for 2005 (H.R. 1268/P.L. 109-13 of May 11, 2005) prohibiting a winner-take-all competition. The provision effectively required the participation of at least one additional shipyard in the program but did not specify the share of the program that is to go to the additional shipyard.

On May 25, 2005, the Navy announced that, in light of Section 1019 of P.L. 109-13, it wanted to shift to a "dual-lead-ship" acquisition strategy, under which two DDG-1000s would be procured in FY2007, with one to be designed and built by HII/Ingalls and the other by GD/BIW.

Section 125 of the FY2006 defense authorization act (H.R. 1815/P.L. 109-163) again prohibited the Navy from using a winner-take-all acquisition strategy for procuring its next-generation destroyer. The provision again effectively requires the participation of at least one additional

---

[47] For more on integrated electric-drive technology, see CRS Report RL30622, *Electric-Drive Propulsion for U.S. Navy Ships: Background and Issues for Congress*, by Ronald O'Rourke.

[48] At the time of the events described in this section, HII was owned by Northrop Grumman and was called Northrop Grumman Shipbuilding (NGSB).

shipyard in the program but does not specify the share of the program that is to go to the additional shipyard.

On November 23, 2005, the USD AT&L granted Milestone B approval for the DDG-1000, permitting the program to enter the System Development and Demonstration (SDD) phase. As part of this decision, the USD AT&L approved the Navy's proposed dual-lead-ship acquisition strategy and a low rate initial production quantity of eight ships (one more than the Navy subsequently planned to procure).

On February 14, 2008, the Navy awarded contract modifications to GD/BIW and HII/Ingalls for the construction of the two lead ships. The awards were modifications to existing contracts that the Navy has with GD/BIW and HII/Ingalls for detailed design and construction of the two lead ships. Under the modified contracts, the line item for the construction of the dual lead ships is treated as a cost plus incentive fee (CPIF) item.

Until July 2007, it was expected that HII/Ingalls would be the final-assembly yard for the first DDG-1000 and that GD/BIW would be the final-assembly yard for the second. On September 25, 2007, the Navy announced that it had decided to build the first DDG-1000 at GD/BIW, and the second at HII/Ingalls.

On January 12, 2009, it was reported that the Navy, HII/Ingalls, and GD/BIW in the fall of 2008 began holding discussions on the idea of having GD/BIW build both the first and second DDG-1000s, in exchange for HII/Ingalls receiving a greater share of the new DDG-51s that would be procured under the Navy's July 2008 proposal to stop DDG-1000 procurement and restart DDG-51 procurement.[49]

On April 8, 2009, it was reported that the Navy had reached an agreement with HII/Ingalls and GD/BIW to shift the second DDG-1000 to GD/BIW, and to have GD/BIW build all three ships. HII/Iingalls will continue to make certain parts of the three ships, notably their composite deckhouses. The agreement to have all three DDG-1000s built at GD/BIW was a condition that Secretary of Defense Robert Gates set forth in an April 6, 2009, news conference on the FY2010 defense budget for his support for continuing with the construction of all three DDG-1000s (rather than proposing the cancellation of the second and third).

Procurement Cost Cap

Section 123 of the FY2006 defense authorization act (H.R. 1815/P.L. 109-163 of January 6, 2006) limited the procurement cost of the fifth DDG-1000 to $2.3 billion, plus adjustments for inflation and other factors. Given the truncation of the DDG-1000 program to three ships, this unit procurement cost cap appears moot.

---

[49] Christopher P. Cavas, "Will Bath Build Second DDG 1000?" *Defense News*, January 12, 2009: 1, 6.

2010 Nunn-McCurdy Breach, Program Restructuring, and Milestone Recertification

On February 1, 2010, the Navy notified Congress that the DDG-1000 program had experienced a critical cost breach under the Nunn-McCurdy provision. The Nunn-McCurdy provision (10 U.S.C. 2433a) requires certain actions to be taken if a major defense acquisition program exceeds (i.e., breaches) certain cost-growth thresholds and is not terminated. Among other things, a program that experiences a cost breach large enough to qualify under the provision as a critical cost breach has its previous acquisition system milestone certification revoked. (In the case of the DDG-1000 program, this was Milestone B.) In addition, for the program to proceed rather than be terminated, DOD must certify certain things, including that the program is essential to national security and that there are no alternatives to the program that will provide acceptable capability to meet the joint military requirement at less cost.[50]

The Navy stated in its February 1, 2010, notification letter that the DDG-1000 program's critical cost breach was a mathematical consequence of the program's truncation to three ships.[51] Since the DDG-1000 program has roughly $9.3 billion in research and development costs, truncating the program to three ships increased to roughly $3.1 billion the average amount of research and development costs that are included in the average acquisition cost (i.e., average research and development cost plus procurement cost) of each DDG-10000. The resulting increase in program acquisition unit cost (PAUC)—one of two measures used under the Nunn-McCurdy provision for measuring cost growth[52]—was enough to cause a Nunn-McCurdy critical cost breach.

In a June 1, 2010, letter (with attachment) to Congress, Ashton Carter, the DOD acquisition executive (i.e., the Under Secretary of Defense for Acquisition, Technology and Logistics), stated that he had restructured the DDG-1000 program and that he was issuing the certifications required under the Nunn-McCurdy provision for the restructured DDG-1000 program to proceed.[53] The letter stated that the restructuring of the DDG-1000 program included the following:

- A change to the DDG-1000's design affecting its primary radar.

- A change in the program's Initial Operational Capability (IOC) from FY2015 to FY2016.

- A revision to the program's testing and evaluation requirements.

Regarding the change to the ship's design affecting its primary radar, the DDG-1000 originally was to have been equipped with a dual-band radar (DBR) consisting of the Raytheon-built X-

---

[50] For more on the Nunn-McCurdy provision, see CRS Report R41293, *The Nunn-McCurdy Act: Background, Analysis, and Issues for Congress*, by Moshe Schwartz.

[51] Source: Letter to congressional offices dated February 1, 2010, from Robert O. Work, Acting Secretary of the Navy, to Representative Ike Skelton, provided to CRS by Navy Office of Legislative Affairs on February 24, 2010.

[52] PAUC is the sum of the program's research and development cost and procurement cost divided by the number of units in the program. The other measure used under the Nunn-McCurdy provision to measure cost growth is average program unit cost (APUC), which is the program's total procurement cost divided by the number of units in the program.

[53] Letter dated June 1, 2010, from Ashton Carter, Under Secretary of Defense (Acquisition, Technology and Logistics) to the Honorable Ike Skelton, with attachment. The letter and attachment were posted on InsideDefense.com (subscription required) on June 2, 2010.

band SPY-3 multifunction radar (MFR) and the Lockheed-built S-band SPY-4 Volume Search Radar (VSR). (Raytheon is the prime contractor for the overall DBR.) Both parts of the DBR have been in development for the past several years. An attachment to the June 1, 2010, letter stated that, as a result of the program's restructuring, the ship is now to be equipped with "an upgraded multifunction radar [MFR] and no volume search radar [VSR]." The change eliminates the Lockheed-built S-band SPY-4 VSR from the ship's design. The ship might retain a space and weight reservation that would permit the VSR to be backfitted to the ship at a later point. The Navy states that

> As part of the Nunn-McCurdy certification process, the Volume Search Radar (VSR) hardware was identified as an acceptable opportunity to reduce cost in the program and thus was removed from the current baseline design....
>
> Modifications will be made to the SPY-3 Multi-Function Radar (MFR) with the focus of meeting ship Key Performance Parameters. The MFR modifications will involve software changes to perform a volume search functionality. Shipboard operators will be able to optimize the SPY-3 MFR for either horizon search or volume search. While optimized for volume search, the horizon search capability is limited. Without the VSR, DDG 1000 is still expected to perform local area air defense....
>
> The removal of the VSR will result in an estimated $300 million net total cost savings for the three-ship class. These savings will be used to offset the program cost increase as a result of the truncation of the program to three ships. The estimated cost of the MFR software modification to provide the volume search capability will be significantly less than the estimated procurement costs for the VSR.[54]

Regarding the figure of $300 million net total cost savings in the above passage, the Navy during 2011 determined that eliminating the SPY-4 VSR from the DDG-1000 increased by $54 million the cost to integrate the dual-band radar into the Navy's new Gerald R. Ford (CVN-78) class aircraft carriers.[55] Subtracting this $54 million cost from the above $300 million savings figure would bring the net total cost savings to about $246 million on a Navy-wide basis.

A July 26, 2010, press report quotes Captain James Syring, the DDG-1000 program manager, as stating: "We don't need the S-band radar to meet our requirements [for the DDG-1000]," and "You can meet [the DDG-1000's operational] requirements with [the] X-band [radar] with software modifications."[56]

An attachment to the June 1, 2010, letter stated that the PAUC for the DDG-1000 program had increased 86%, triggering the Nunn-McCurdy critical cost breach, and that the truncation of the program to three ships was responsible for 79 of the 86 percentage points of increase. (The attachment stated that the other seven percentage points of increase are from increases in development costs that are primarily due to increased research and development work content for the program.)

---

[54] Source: Undated Navy information paper on DDG-51 program restructuring provided to CRS and CBO by Navy Office of Legislative Affairs on July 19, 2010.

[55] Source: Undated Navy information paper on CVN-78 cost issues, provided by Navy Office of Legislative Affairs to CRS on March 19, 2012.

[56] Cid Standifer, "Volume Radar Contracted For DDG-1000 Could Be Shifted To CVN-79," *Inside the Navy*, July 26, 2010.

Carter also stated in his June 1, 2010, letter that he had directed that the DDG-1000 program be funded, for the period FY2011-FY2015, to the cost estimate for the program provided by the Cost Assessment and Program Evaluation (CAPE) office (which is a part of the Office of the Secretary of Defense [OSD]), and, for FY2016 and beyond, to the Navy's cost estimate for the program. The program was previously funded to the Navy's cost estimate for all years. Since CAPE's cost estimate for the program is higher than the Navy's cost estimate, funding the program to the CAPE estimate for the period FY2011-FY2015 will increase the cost of the program as it appears in the budget for those years. The letter states that DOD "intends to address the [resulting] FY2011 [funding] shortfall [for the DDG-1000 program] through reprogramming actions."

An attachment to the letter stated that the CAPE in May 2010 estimated the PAUC of the DDG-1000 program (i.e., the sum of the program's research and development costs and procurement costs, divided by the three ships in the program) as $7.4 billion per ship in then-year dollars ($22.1 billion in then-year dollars for all three ships), and the program's average procurement unit cost (APUC), which is the program's total procurement cost divided by the three ships in the program, as $4.3 billion per ship in then-year dollars ($12.8 billion in then-year dollars for all three ships). The attachment stated that these estimates are at a confidence level of about 50%, meaning that the CAPE believes there is a roughly 50% chance that the program can be completed at or under these cost estimates, and a roughly 50% chance that the program will exceed these cost estimates.

An attachment to the letter directed the Navy to "return for a Defense Acquisition Board (DAB) review in the fall 2010 timeframe when the program is ready to seek approval of the new Milestone B and authorization for production of the DDG-1002 [i.e., the third ship in the program]."

On October 8, 2010, DOD reinstated the DDG-1000 program's Milestone B certification and authorized the Navy to continue production of the first and second DDG-1000s and commence production of the third DDG-1000.[57]

Under Secretary of Defense Ashton Carter's June 1, 2010, letter and attachment restructuring the DDG-1000 program and DOD's decision on October 8, 2010, to reinstate the DDG-1000 program's Milestone B certification (see **Appendix A**) raise the following potential oversight questions for Congress:

- Why did DOD decide, as part of its restructuring of the DDG-1000 program, to change the primary radar on the DDG-1000?

- What are the potential risks to the DDG-1000 program of changing its primary radar at this stage in the program (i.e., with the first ship under construction, and preliminary construction activities underway on the second ship)?

- How will the upgraded MFR differ in cost, capabilities, and technical risks from the baseline MFR included in the original DDG-1000 design?

- What is the net impact on the capabilities of the DDG-1000 of the change to the DDG-1000's primary radar (i.e., of removing the VSR and upgrading the MFR)?

---

[57] Christopher J. Castelli, "Pentagon Approves Key Milestone For Multibillion-Dollar Destroyer," *Inside the Navy*, November 22, 2010.

- Given change to the DDG-1000's primary radar and the May 2010 CAPE estimates of the program's program acquisition unit cost (PAUC) and average program unit cost (APUC), is the DDG-1000 program still cost effective?

- What impact on cost, schedule, or technical risk, if any, will the removal of the VSR from the DDG-1000 design have on the Navy's plan to install the dual-band radar (DBR), including the VSR, on the Ford (CVN-78) class aircraft carriers CVN-78 and CVN-79?[58]

March 2012 GAO Report

A March 2012 GAO report assessing major DOD weapon acquisition programs stated the following of the DDG-1000 program:

Technology Maturity

The DDG 1000 program is still conducting development work on several of its critical technologies. Three of DDG 1000's 12 critical technologies are currently mature and the integrated deckhouse will be delivered to the first ship for installation in fiscal year 2012. However, the remaining eight technologies will not be demonstrated in a realistic environment until after ship installation. The Navy has completed a successful full-power test of the integrated power system—one of these eight technologies—and plans a follow-on test in fiscal year 2012. Another critical technology—the total ship computing environment—consisted initially of six software releases. According to program officials, software release 5 has been completed and was used in land-based testing in fiscal year 2011. The program has made changes to release 6, and has prioritized the software needed to support shipyard delivery over the functionality needed for activating the mission systems. This functionality was moved out of the releases and will be developed as part of a spiral. Other key technologies, including the multifunction radar and the advanced gun system, have been delivered to the first ship. However, the gun system's long-range land-attack projectile has encountered delays, primarily due to problems with its rocket motor. The Navy plans to finalize and test the rocket motor design by March 2012. The Navy has performed several guided flight tests using older rocket motor designs, which demonstrated that the projectile can meet its accuracy and range requirements.

Design and Production Maturity

The DDG 1000 design is stable, although the potential for design changes remains until its critical technologies are fully mature. There have been few design changes resulting from lead-ship production, which program officials attribute to a high level of design maturity prior to fabrication. As of January 2012, the Navy estimated that the first ship, which began fabrication in February 2009, was 63 percent complete. The second ship, which began fabrication in March 2010, is 22 percent complete.

Shipbuilders have experienced several challenges in constructing the first and second ships, including issues with the manufacture and installation of certain composite materials. According to program officials, the shipbuilder had to rework some areas, which resulted in an 8-week schedule delay, but only minor cost increases. Program officials noted that the

---

[58] For more on these aircraft carriers, see CRS Report RS20643, *Navy Ford (CVN-78) Class Aircraft Carrier Program: Background and Issues for Congress*, by Ronald O'Rourke.

shipbuilder has developed and refined manufacturing and installation processes to reduce the likelihood of rework on the subsequent ships.

Other Program Issues

Following a Nunn-McCurdy unit-cost breach of the critical threshold in 2010, DOD restructured the DDG 1000 program and removed the volume-search radar from the ship's design. The program will modify the multifunction radar to meet its volume-search requirements.

The Navy recently awarded contracts for fabrication of the third ship, its gun system, and software development and integration. In these negotiations, program officials said the Navy leveraged its knowledge about actual costs from the fabrication of the first two ships and from other programs to reduce costs. According to program officials, the Navy is in negotiations for contracts related to the mission systems and other key segments of the third ship.

Program Office Comments

In commenting on a draft of this assessment, Navy officials confirmed that the program was recertified and restructured in June 2010 following a Nunn-McCurdy unit-cost breach and the curtailing of the program to three ships. Since then, the Navy has awarded ship construction contracts for all three ships, including advanced guns and all required software. All critical technologies have been tested in at least a relevant environment. Officials noted that more than 90 percent of the software has been completed through design, code, unit test, and integration and its schedule is aligned to ship activations. Navy officials also said that program reviews by the Office of the Under Secretary of Defense for Acquisition, Technology and Logistics and DOD's Office of Performance Assessment and Root Cause Analyses concluded that the Navy is managing and retiring program risk. The Navy also provided technical comments, which were incorporated as appropriate.[59]

---

[59] Government Accountability Office, *Defense Acquisitions[:] Assessments of Selected Weapon Programs*, GAO-12-400SP, March 2012, p. 68.

Appendix B. Additional Background Information on CG(X) Cruiser Program

Background Information on CG(X) Program

The CG(X) cruiser program was announced by the Navy on November 1, 2001.[60] The Navy wanted to procure as many as 19 CG(X)s as replacements for its 22 CG-47s, which are projected to reach the end of their 35-year service lives between 2021 and 2029. The CG-47s are multi-mission ships with an emphasis on AAW and (for some CG-47s) BMD, and the Navy similarly wanted the CG(X) to be a multi-mission ship with an emphasis on AAW and BMD. The CG(X) was to carry the Air and Missile Defense Radar (AMDR), a new radar that was to be considerably larger and more powerful than the SPY-1 radar carried on the Navy's Aegis ships. Some press reports suggested that a nuclear-powered version of the CG(X) might have had a full load displacement of more than 20,000 tons and a unit procurement cost of $5 billion or more.[61]

The Navy's FY2009 budget called for procuring the first CG(X) in FY2011. Beginning in late 2008, however, it was reported that the Navy had decided to defer the procurement of the first CG(X) by several years, to about FY2017.[62] Consistent with these press reports, on April 6, 2009,

---

[60] The Navy on that date announced that that it was launching a Future Surface Combatant Program aimed at acquiring a family of next-generation surface combatants. This new family of surface combatants, the Navy stated, would include three new classes of ships:

- a destroyer called the DD(X)—later redesignated DDG-1000—for the precision long-range strike and naval gunfire mission,
- a cruiser called the CG(X) for the AAW and BMD mission, and
- a smaller combatant called the Littoral Combat Ship (LCS) to counter submarines, small surface attack craft, and mines in heavily contested littoral (near-shore) areas.

The Future Surface Combatant Program replaced an earlier Navy surface combatant acquisition effort, begun in the mid-1990s, called the Surface Combatant for the 21st Century (SC-21) program. The SC-21 program encompassed a planned destroyer called DD-21 and a planned cruiser called CG-21. When the Navy announced the Future Surface Combatant Program in 2001, development work on the DD-21 had been underway for several years, but the start of development work on the CG-21 was still years in the future. The DD(X) program, now called the DDG-1000 or Zumwalt-class program, is essentially a restructured continuation of the DD-21 program. The CG(X) might be considered the successor, in planning terms, of the CG-21. After November 1, 2001, the acronym SC-21 continued for a time to be used in the Navy's research and development account to designate a line item (i.e., program element) that funded development work on the DDG-1000 and CG(X).

[61] For a discussion of nuclear power for Navy surface ships other than aircraft carriers, see CRS Report RL33946, *Navy Nuclear-Powered Surface Ships: Background, Issues, and Options for Congress*, by Ronald O'Rourke.

[62] Zachary M. Peterson, "Navy Awards Technology Company $128 Million Contract For CG(X) Work," *Inside the Navy*, October 27, 2008. Another press report (Katherine McIntire Peters, "Navy's Top Officer Sees Lessons in Shipbuilding Program Failures," *GovernmentExecutive.com*, September 24, 2008) quoted Admiral Gary Roughead, the Chief of Naval Operations, as saying: "What we will be able to do is take the technology from the DDG-1000, the capability and capacity that [will be achieved] as we build more DDG-51s, and [bring those] together around 2017 in a replacement ship for our cruisers." (Material in brackets in the press report.) Another press report (Zachary M. Peterson, "Part One of Overdue CG(X) AOA Sent to OSD, Second Part Coming Soon," *Inside the Navy*, September 29, 2008) quoted Vice Admiral Barry McCullough, the Deputy Chief of Naval Operations for Integration of Capabilities and Resources, as saying that the Navy did not budget for a CG(X) hull in its proposal for the Navy's budget under the FY2010-FY2015 Future Years Defense Plan (FYDP) to be submitted to Congress in early 2009.

An earlier report (Christopher P. Cavas, "DDG 1000 Destroyer Program Facing Major Cuts," *DefenseNews.com*, July 14, 2008) stated that the CG(X) would be delayed until FY2015 or later. See also Geoff Fein, "Navy Likely To Change (continued...)

Secretary of Defense Robert Gates announced—as part of a series of recommendations for the then-forthcoming FY2010 defense budget—a recommendation to "delay the CG-X next generation cruiser program to revisit both the requirements and acquisition strategy" for the program.[63] The Navy's proposed FY2010 budget deferred procurement of the first CG(X) beyond FY2015.

Cancellation of CG(X) Program

The Navy's FY2011 budget proposed terminating the CG(X) program as unaffordable. The Navy's desire to cancel the CG(X) and instead procure Flight III DDG-51s apparently took shape during 2009: at a June 16, 2009, hearing before the Seapower Subcommittee of the Senate Armed Services Committee, the Navy testified that it was conducting a study on destroyer procurement options for FY2012 and beyond that was examining design options based on either the DDG-51 or DDG-1000 hull form.[64] A January 2009 memorandum from the Department of Defense acquisition executive had called for such a study.[65] In September and November 2009, it was reported that the Navy's study was examining how future requirements for AAW and BMD operations might be met by a DDG-51 or DDG-1000 hull equipped with a new radar.[66] On December 7, 2009, it was reported that the Navy wanted to cancel its planned CG(X) cruiser and instead procure an improved version of the DDG-51.[67] In addition to being concerned about the projected high cost and immature technologies of the CG(X),[68] the Navy reportedly had concluded that it does not need a surface combatant with a version of the AMDR as large and capable as the one envisaged for the CG(X) to adequately perform projected AAW and BMD missions, because the Navy will be able to augment data collected by surface combatant radars with data collected by space-based sensors. The Navy reportedly concluded that using data collected by other sensors would permit projected AAW and BMD missions to be performed

---

(...continued)

CG(X)'s Procurement Schedule, Official Says," *Defense Daily*, June 24, 2008; Rebekah Gordon, "Navy Agrees CG(X) By FY-11 Won't Happen But Reveals Little Else," *Inside the Navy*, June 30, 2008.

[63] Source: Opening remarks of Secretary of Defense Robert Gates at an April 6, 2009, news conference on DOD recommendations for the then-forthcoming FY2010 defense budget.

[64] Source: Transcript of spoken remarks of Vice Admiral Bernard McCullough at a June 16, 2009, hearing on Navy force structure shipbuilding before the Seapower subcommittee of the Senate Armed Services Committee.

[65] A January 26, 2009, memorandum for the record from John Young, the then-DOD acquisition executive, stated that "The Navy proposed and OSD [the Office of the Secretary of Defense] agreed with modification to truncate the DDG-1000 Program to three ships in the FY 2010 budget submission." The memo proposed procuring one DDG-51 in FY2010 and two more FY2011, followed by the procurement in FY2012-FY2015 (in annual quantities of 1, 2, 1, 2) of a ship called the Future Surface Combatant (FSC) that could be based on either the DDG-51 design or the DDG-1000 design. The memorandum stated that the FSC might be equipped with a new type of radar, but the memorandum did not otherwise specify the FSC's capabilities. The memorandum stated that further analysis would support a decision on whether to base the FSC on the DDG-51 design or the DDG-1000 design. (Memorandum for the record dated January 26, 2009, from John Young, Under Secretary of Defense [Acquisition, Technology and Logistics], entitled "DDG 1000 Program Way Ahead," posted on InsideDefense.com [subscription required].)

[66] Zachary M. Peterson, "Navy Slated To Wrap Up Future Destroyer Hull And Radar Study," *Inside the Navy*, September 7, 2009. Christopher P. Cavas, "Next-Generation U.S. Warship Could Be Taking Shape," *Defense News*, November 2, 2009: 18, 20.

[67] Christopher J. Castelli, "Draft Shipbuilding Report Reveals Navy Is Killing CG(X) Cruiser Program," *Inside the Navy*, December 7, 2009.

[68] Christopher J. Castelli, "Draft Shipbuilding Report Reveals Navy Is Killing CG(X) Cruiser Program," *Inside the Navy*, December 7, 2009.

---

adequately with a radar smaller enough to be fitted onto the DDG-51.[69] Reports suggested that the new smaller radar would be a scaled-down version of the AMDR originally intended for the CG(X).[70]

The Navy's February 2010 report on its FY2011 30-year (FY2011-FY2040) shipbuilding plan, submitted to Congress in conjunction with the FY2011 budget, states that the 30-year plan:

> Solidifies the DoN's [Department of the Navy's] long-term plans for Large Surface Combatants by truncating the DDG 1000 program, restarting the DDG 51 production line, and continuing the Advanced Missile Defense Radar (AMDR) development efforts. Over the past year, the Navy has conducted a study that concludes a DDG 51 hull form with an AMDR suite is the most cost-effective solution to fleet air and missile defense requirements over the near to mid-term....
>
> The Navy, in consultation with OSD, conducted a Radar/Hull Study for future destroyers. The objective of the study was to provide a recommendation for the total ship system solution required to provide Integrated Air and Missile Defense (IAMD) (simultaneous ballistic missile and anti-air warfare (AAW) defense) capability while balancing affordability with capacity. As a result of the study, the Navy is proceeding with the Air and Missile Defense Radar (AMDR) program....
>
> As discussed above, the DDG 51 production line has been restarted. While all of these new-start guided missile destroyers will be delivered with some BMD capability, those procured in FY 2016 and beyond will be purpose-built with BMD as a primary mission. While there is work to be done in determining its final design, it is envisioned that this DDG 51 class variant will have upgrades to radar and computing performance with the appropriate power generation capacity and cooling required by these enhancements. These upgraded DDG 51 class ships will be modifications of the current guided missile destroyer design that combine the best emerging technologies aimed at further increasing capabilities in the IAMD arena and providing a more effective bridge between today's capability and that originally planned for the CG(X). The ships reflected in this program have been priced based on continuation of the existing DDG 51 re-start program. Having recently completed the Hull and Radar Study, the Department is embarking on the requirements definition process for these AMDR destroyers and will adjust the pricing for these ships in future reports should that prove necessary.[71]

In testimony to the House and Senate Armed Services Committees on February 24 and 25, 2010, respectively, Admiral Gary Roughead, the Chief of Naval Operations, stated:

> Integrated Air and Missile Defense (IAMD) incorporates all aspects of air defense against ballistic, anti-ship, and overland cruise missiles. IAMD is vital to the protection of our force, and it is an integral part of our core capability to deter aggression through conventional means....

---

[69] Amy Butler, "STSS Prompts Shift in CG(X) Plans," *Aerospace Daily & Defense Report*, December 11, 2009: 1-2.

[70] Cid Standifer, "NAVSEA Plans To Solicit Contracts For Air And Missile Defense Radar," *Inside the Navy*, December 28, 2009; "Navy Issues RFP For Phase II of Air And Missile Defense Radar Effort," *Defense Daily*, December 24, 2009: 4.

[71] U.S. Navy, *Report to Congress on Annual Long-Range Plan for Construction of Naval Vessels for FY 2011*, February 2010, pp. 12, 13, 19. The first reprinted paragraph, taken from page 12, also occurs on page 3 as part of the executive summary.

To address the rapid proliferation of ballistic and anti-ship missiles and deep-water submarine threats, as well as increase the capacity of our multipurpose surface ships, we restarted production of our DDG 51 Arleigh Burke Class destroyers (Flight IIA series). These ships will be the first constructed with IAMD, providing much-needed Ballistic Missile Defense (BMD) capacity to the Fleet, and they will incorporate the hull, mechanical, and electrical alterations associated with our mature DDG modernization program. We will spiral DDG 51 production to incorporate future integrated air and missile defense capabilities....

The Navy, in consultation with the Office of the Secretary of Defense, conducted a Radar/Hull Study for future surface combatants that analyzed the total ship system solution necessary to meet our IAMD requirements while balancing affordability and capacity in our surface Fleet. The study concluded that Navy should integrate the Air and Missile Defense Radar program S Band radar (AMDR-S), SPY-3 (X Band radar), and Aegis Advanced Capability Build (ACB) combat system into a DDG 51 hull. While our Radar/Hull Study indicated that both DDG 51 and DDG 1000 were able to support our preferred radar systems, leveraging the DDG 51 hull was the most affordable option. Accordingly, our FY 2011 budget cancels the next generation cruiser program due to projected high cost and risk in technology and design of this ship. I request your support as we invest in spiraling the capabilities of our DDG 51 Class from our Flight IIA Arleigh Burke ships to Flight III ships, which will be our future IAMD-capable surface combatant. We will procure the first Flight III ship in FY 2016.[72]

Author Contact Information

Ronald O'Rourke
Specialist in Naval Affairs
rorourke@crs.loc.gov, 7-7610

---

[72] Statement of Admiral Gary Roughead, Chief of Naval Operations, before the House Armed Services Committee on 24 February, 2010, pp. 10-11; and Statement of Admiral Gary Roughead, Chief of Naval Operations, before the Senate Armed Services Committee on 25 February 2010, pp. 10-11.

www.ingramcontent.com/pod-product-compliance
Lightning Source LLC
Chambersburg PA
CBHW081613170526
45166CB00009B/2941